Flow Control Through Bio-inspired Leading-Edge Tubercles

Daniel T. H. New · Bing Feng Ng

Editors

Flow Control Through Bio-inspired Leading-Edge Tubercles

Morphology, Aerodynamics, Hydrodynamics and Applications

 Springer

Editors
Daniel T. H. New
School of Mechanical and Aerospace
Engineering
Nanyang Technological University
Singapore, Singapore

Bing Feng Ng
School of Mechanical and Aerospace
Engineering
Nanyang Technological University
Singapore, Singapore

ISBN 978-3-030-23794-3   ISBN 978-3-030-23792-9 (eBook)
https://doi.org/10.1007/978-3-030-23792-9

This Springer imprint is published by the registered company Springer Nature Switzerland AG
The registered company address is: Gewerbestrasse 11, 6330 Cham, Switzerland

This book is dedicated to our mentors and friends, as well as past and present postdoctoral researchers and graduate students, who had given us so much.

Foreword

The field of biomimetics attempts to use nature as a resource to transition anatomical and physiological features from biological organisms into engineered systems. The expectation is that the lessons learned from biology can improve the performance of engineered systems. Evolution has been responsible for a system of natural experimentation in which organisms performed the cost-benefit analysis through the Darwinian mechanism of natural selection. The multitude of species, both extant and extinct, had to evolve adaptations as solutions to problems related to survival in their biological and physical environments. Many of these problems are the same that humans attempt to solve through their own technologies. Thus, biomimetics provides a new way of thinking and inspiration so that living organisms become a wide resource base of adaptations, which can be adapted for engineering applications. Emulation of organisms through the manufacture of devices has been a long-standing idea both in mythology and history. Icarus and Daedalus were supposed to have escaped their imprisonment by fashioning bird-like wings for flight. Birds as model systems for flying machines were incorporated into the designs of Da Vinci, Lilienthal, and the Wright brothers. Borelli (circa 1680) designed a submarine with buoyancy control based on the swim bladder of a fish and a propulsion system inspired by the action of the paddling feet of frogs.

In more recent times, a partnership between biologists and engineers has focused attention on the use of biological organisms as model systems to improve engineered systems. Examples of biomimetic applications have included Velcro from plant burrs, drag reduction from shark scales, concept car design based on the boxfish, dry adhesion based on the gecko foot, air conditioning from examination of termite mounds, structural color and its application to paints and cosmetics, flapping fins and wings for bio-inspired aquatic and aerial drones, the lotus leaf effect, spider silk, bullet train design based on the beak of kingfishers, exoskeletons, cooperative behavior, and echolocation, to mention a few areas of potential application. While biomimetic systems continue to be proposed, many are still at an early stage of development and although extremely tantalizing they have been elusive in an understanding of the physics of these systems and their eventual application for commercial products.

The leading-edge tubercles of the flipper of the humpback whale have proved to be a structural adaptation that affords the whale enhanced maneuvering performance. In elucidating the physics behind the function of the tubercles, this research has opened up new avenues in aero/hydrodynamic flow control and provided new commercial opportunities. While the original discovery in the biology of the structural design and mechanics was serendipitous, the purpose of the tubercles was the result of hydrodynamic analysis and testing from engineering. What has been ascertained is that the tubercles control flow over wing-like surfaces. By generating paired vortices in what can be called a sacrificed separation, the tubercles can maintain regions of accelerated flow along a wing that avoid separation and delay the stall condition beyond what would be normal for wings devoid of tubercles. This form of bio-inspired passive flow control has potential applications in aircraft and nautical design, wind and water turbines, compressors, and pumps. The tubercles on whale flippers also represent an authentic example for biomimetics that has inspired future generations of students to pursue careers in biology, engineering, and energy conservation.

This book offers a snapshot of the research on leading-edge tubercles as more research on the phenomenon is being performed around the world. The breadth of the chapters covered in this book demonstrated the significance and interest on the topic. A leading expert in the field of biology or engineering wrote each chapter. The topics covered address key issues pertaining to the function of the leading-edge tubercles. Chapter "Biomimetics and the Application of the Leading-Edge Tubercles of the Humpback Whale Flipper" defines the biomimetic approach, provides an overview of the biology of the humpback whale with respect to the function of the tubercles, and reviews the potential applications. Chapter "Tubercled Wing Flow Physics and Performance" discusses wind tunnel experiments focusing on the aerodynamic properties involving the tubercles as vortex generators, their Reynolds number effects, and experimentation in aeroacoustics. In chapter "Tubercle Geometric Configurations: Optimization and Alternatives," the geometry of the tubercles for optimization of aerodynamic performance is explored including alternative designs affecting drag. In chapter "Flow Control by Hydrofoils with Leading-Edge Tubercles," flow behavior and control are examined from water tunnel studies and computational fluid dynamics demonstrating the tubercles as vortex generators and exploring their function in low Reynolds number flow. The effect of tubercles on spanwise flow is examined in chapter "Leading-Edge Tubercles on Swept and Delta Wing Configurations" with an emphasis on swept and delta wing configurations. Chapter "Effects of Leading-Edge Tubercles on Dynamically Pitching Airfoils" introduces the reader to actively pitching airfoils and how leading-edge tubercles control dynamic stall. Lastly, the aeroelastic effects with structural considerations are reviewed in chapter "Effects of Leading-Edge Tubercles on Structural Dynamics and Aeroelasticity".

While much of the scientific aspects of the leading-edge tubercles that are presented in this book are geared toward engineering, this book has utility for biologists as well in demonstrating how engineering techniques can be used to study the biological role of adaptations possessed by an organism. In addition, biologists can see how particular adaptations can be selected that exhibit the novel qualities that make them appropriate for biomimetic applications. As more sophisticated methodologies are developed to contend with the particular attributes of living organisms (i.e., small size, chemical motors, compliant wet structural materials, complex neural networks, self-healing), greater union among biologists and engineers will flourish to advance both fields through biomimetics and bio-inspired design.

Frank E. Fish
Liquid Life Laboratory, Department of Biology
West Chester University
West Chester, PA, USA

Preface

Appreciating and leveraging upon Mother Nature's well-honed solutions, or bio-mimicry, have always been behind many engineering innovations. From early flying machines envisioned by Da Vinci, shark-skin-inspired surfaces for drag reduction, to bat echolocation for target tracking optimizations, humans have always attempted to adapt nature's answers to physical challenges towards solving engineering problems and create new solutions. Over the last two decades since protuberances or tubercles along whale-flipper leading-edges (LE) had been postulated to be responsible for the underwater maneuverability of humpback whales, a tremendous amount of research efforts had been expended to elucidate the capabilities of LE tubercles in terms of their functionalities, flow mechanisms underlying favorable flow influences, and how they may be optimized for various hydrodynamics and aerodynamics applications. It is now a generally accepted notion that the implementation of LE tubercles upon lifting surfaces with the appropriate geometries tends to delay stall through mitigation of flow separation events, as compared to conventional lifting surfaces that do not have any. The unique physical features of LE tubercled lifting surfaces also led to highly three-dimensional and intriguing flow features, especially when different surface planforms such as swept or delta configurations are used. Several studies have reported that the LE tubercles effectively function as vortex generators, while some reported them as influencing both the streamwise and spanwise pressure gradients. Regardless, good flow control abilities of appropriately configured LE tubercles are observed and recognized in most experimental and simulation studies. In fact, the usefulness of leading-edge tubercles goes beyond conventional planar lifting surfaces, as several studies have demonstrated their potential usage upon marine propellers and wind turbine blades as well. Furthermore, their effectiveness also extends beyond simply increasing lift and reducing drag, as studies have shown that they may be able to reduce wing flutter.

As the study of LE tubercles gradually transits from understanding them and their flow effects to implementing them for proof-of-concepts or actual engineering applications, their designs and optimizations have also become increasingly complex. While the use of modern experimental techniques and numerical flow modeling helped to mitigate some of these challenges, a firm grasp of the fundamentals and the preceding research work associated with LE tubercles remains necessary, particularly in the selection and optimization of LE tubercle geometries/configurations for specific usages. As such, a book that provides key information upon the morphology of LE tubercles, how earlier studies have better inform researchers in their design decisions, flow characteristics, and performance, as well as some of the possible applications that they could be used for, will be timely and useful to researchers at this point in time. The editors and authors hope that this book will expose new readers to this research area and inform them of the current state-of-the-art research and application information, as well as inspiring them to tap into the knowledge that has been garnered thus far. To do that, this book will first introduce the readers to the practice of biomimetics with a particular emphasis on humpback whale flipper, as well as some of the new emerging engineering applications that tap into this novel bio-inspired solution. Thereafter, aerodynamics and hydrodynamics performance associated with LE tubercled wings and hydrofoils will be elaborated. Subsequently, the effects of significant spanwise flow changes induced by the LE tubercles on swept and delta wings will be discussed, followed by the flow behavior of LE tubercled wings undergoing oscillatory pitching motions. To wrap things up, new insights on the potential use of LE tubercles to control wing flutter will be presented.

Last but not least, the editors would like to express their deep appreciation toward all authors for their contributions toward the present book despite their busy schedules. It would not have been possible without their dedication and enthusiasm.

<table>
<tr><td>Singapore</td><td>Daniel T. H. New</td></tr>
<tr><td>April 2019</td><td>Bing Feng Ng</td></tr>
</table>

Contents

Biomimetics and the Application of the Leading-Edge Tubercles of the Humpback Whale Flipper

Frank E. Fish

Abstract The field of biomimetics attempts to inspire and integrate the morphology and function of biological organisms into the design of human-made technology. In that organisms have been able to adapt through evolution, they have performed the "cost-benefit analysis" to solve a variety of problems of concern to humans and can potentially improve technologies. One example of a structural adaptation that can improve the aero/hydrodynamic performance of wing-like designs is based on the flippers of the humpback whale. The humpback whale is able produce small radius turns with its elongate, high aspect ratio flippers. This whale differs from related species in using maneuverability to capture prey. Maintenance of lift throughout a turning maneuver requires a modification of the wing-like flippers. The flippers possess rounded bumps, called tubercles, along the leading-edge. Empirical and computational studies have demonstrated that the tubercles passively modify the flow over wing-like structures. The flow between the tubercles produces counter-rotating vortices in a sacrificed separation that helps to energize the flow over the tubercles. The flow pattern over a wing induced by the tubercles increases lift, delays stall, and maintains low drag post stall. The tubercles have applications for aircraft wings, rudders, dive planes, skegs, sailboat masts, stabilizers, truck mirrors, bicycle wheels, rotor blades, propellers, compressors, pumps, fans, and tidal and wind turbines.

Keywords Tubercles · Humpback whale · Stall · Lift · Biomimicry · Bio-inspired design

1 Introduction

Since Prometheus brought fire to humanity and set off a technological revolution, humans have attempted with their newfound knowledge to control and conquer the nature. The development of technologies and the advancement of civilization have even attempted to divorce us further from the natural world (Burke 1978). We have

F. E. Fish (✉)
Department of Biology, Liquid Life Laboratory, West Chester University, West Chester, PA 19383, USA
e-mail: ffish@wcupa.edu

© Springer Nature Switzerland AG 2020
D. T. H. New and B. F. Ng (eds.), *Flow Control Through Bio-inspired Leading-Edge Tubercles*, https://doi.org/10.1007/978-3-030-23792-9_1

learned to substitute wheels for legs, concrete and steel for bone and wood, and nylon and Naugahyde for fur and leather. Indeed, with the use of new materials and advanced technologies, we as a species sometimes have surpassed nature. Jet airplanes fly faster than even the swiftest bird. Cars and trains travel at speeds unparalleled even by the cheetah. Our own brains do not have the computation muscle of the new supercomputers.

Any attempt to sever the bonds between humans with their technology and nature can be shortsighted. Nature can serve as a rich source of novel inventions. It has been a long-standing idea that new technologies can be inspired from nature (Benyus 1997; Fish 1998, 2006; Vogel 1998; Lu 2004; Bar-Cohen 2006; Bhushan 2009; Fish and Beneski 2014; Gravish and Lauder 2018). Animals and plants have served as the inspiration for various technological developments. Velcro is one of the most famous products that was inspired from the cockleburs found in nature (Velcro 1955; Bhushan 2009). Drag reduction by riblets was inspired by the scales of sharks (Walsh 1990; Bechert et al. 2000; Oeffner and Lauder 2012; Afroz et al. 2017). Would humans have considered flight without the example of birds and insects (Lillienthal 1911; Wegener 1991; Anderson 1998; McCullough 2015)? Indeed, new research efforts and products and ideas are being inspired from nature in the field of biomimetics (Bar-Cohen 2006; Bhushan 2009; Allen 2010; Faber et al. 2018).

One particular biomimetic device, the tubercles of the leading-edge (LE) of flippers of the humpback whale, has captured the attention of industry and the public (Ingram 2008; Mueller 2008; Quinn and Gaughran 2010; Harman 2013; Kahn 2017). The tubercles represent a simple design feature on the LE of wing-like structures that can beneficially modify the flow over a wing and improve aero-hydrodynamic performance. The purpose of this chapter is to provide introductions to the field of biomimetics and the bio-inspired tubercle technology. The chapter has been written for both biologists and engineers. In this way, the practitioners of these two distinct disciplines can be brought together to appreciate the origins, mechanics and applications of a novel biomimetic design. By examining the biology and physics of tubercles with their advantages, limitations, and constraints, this nature-based technology can serve as an inspiration for the expanded development of this idea and stimulate the development of other biomimetic products.

2　Biomimetics

Nature can be considered as a template to improve mechanical devices and operations, and as the inspiration to develop whole new technologies (Benyus 1997; Vogel 1998; Forbes 2005; Bar-Cohen 2006; Mueller 2008; Allen 2010). An important source of innovative ideas has come from the fields of biomimetics and bio-inspired design. Biomimetics is the field of study that attempts to incorporate novel structures and mechanisms inspired by nature into the design and operation of human-based technologies. The biomimetic approach seeks technological advancement through a transfer of innovation from natural to engineered systems. The goal of biomimetics is to produce engineered systems that possess characteristics of, resemble, or function like living systems, particularly in instances where the performance and morphology of an animal exceeds current mechanical technology (Vogel 1998). Exploiting nature by use of the biomimetic approach or bio-inspiration attempts to seek common solutions from biology and engineering that allow for increased efficiency and specialization (Vincent 1990; Ralston and Swain 2009). By emulating biological characteristics in those instances where the performance of organisms excels manufactured devices, the performance of engineered systems may be improved through biomimetics. Furthermore, engineers can target the diverse morphological and physiological specializations exhibited in biology for technology transfer and effectively reduce the time in the development of innovative technological solutions.

Biological organisms are living machines and subject to the same physical laws as engineered systems. These biological machines are fully autonomous, self-powered, and self-repairing. The component structures that comprise these entities are adapted for particular functions that allow the organism to occupy a specialized environmental functional space, or niche. The multitude of biological designs are the result of the evolutionary process of "natural selection" (Darwin 1859), in which biological organisms can be considered to have performed the "cost-benefit-analysis", optimizing particular designs for specific functions. Biology thus has provided a design prototype (Allen 2010). Over eons, a variety of lineages of organisms have, in effect, experimented with various design characteristics to enhance performance for survival and reproductive success. Thus, the planet can be considered an enormous natural laboratory where an infinite number of experimental trials have been attempted over millions of years (Lu 2004; Bar-Cohen 2012). The results of these experiments are the varied assemblages of living organisms that have inhabited the Earth today and in the past.

It is enticing to contemplate the development of new technological designs for enhanced performance based on living systems. However, such innovations have been elusive (Fish 1998, 2006). The commercial production of products based on biomimicry has been rare (Vogel 1998). There was greater success from bio-inspiration. Strict adherence to biological designs in biomimicry rarely produces any practical results (Vogel 1994, 1998; Fish 1998). It has even been argued that in some cases biomimicry can impede the advancement of engineered systems (Harris 1989; Vogel 1998). Bioinspiration selects ideas and takes into consideration differences between engineered and biological systems by expanding and improving the original biological concept (Ralston and Swain 2009).

'Cultural' differences between biologists and engineers can hamper the creation of biomimetic or bio-inspired technologies (Fish and Beneski 2014). A point of view that differs between biologists and engineers centers around the concept of 'design'. For biologists, design refers to a description of the physical structure of a component or a whole organism in relation to the environment for which it must interact. In this sense, design implies only a functionally proficient arrangement of the parts composing an organism; these components are the result of natural selection (Vogel 1988). Therefore, use of the term design from a biological perspective is simply an indication of the linkage between the structure and function of a characteristic possessed by an organism (Fish 2000). For biologists, design does not infer construction or organization of an organism's characteristics toward a specific goal (Gosline 1991). For the engineer, the concept of design encompasses both the structure of a system and the process by which the system is conceptualized for a particular function. Under this consideration, design is a human endeavor that implies anticipation and purpose. The design process involves discovery, planning, development, construction, evaluation, and invention (Vogel 1998).

There are limits to the biomimetic approach (Fish 2006; Fish and Beneski 2014). Differences between engineered systems and animal systems are apparent. Organisms operate with a different set of principles than those of human-designed technologies (Forbes 2005). Engineered systems are composed of rigid materials, are relatively large in size, use rotation motors, are controlled by computational systems with restricted feedback controls with limited sensory input, are designed for a single function, require maintenance, and are replicated by a manufacturing process that avoids variation. On the other hand, biological organisms are generally small in size, are composed of compliant wet materials, move with rotational motors confined to the microscopic level or with translational movements produced by muscles, have complex neural networks with multiple sensory inputs, are multitasking entities (i.e., they move, feed, and reproduce), are self-healing, and must reproduce in a process that enhances variation (Fish and Beneski 2014).

Perhaps the greatest difference between biological organisms and engineered systems is that engineered systems are a product of conscience design and manufacturing, whereas organisms are the result of evolution (descent with modification) that is fostered and enhanced by the Darwinian process of natural selection. Evolution is not conscience. Evolution has nurtured improvements in design through adaptations for the survival of these organisms by enhanced levels of performance. Natural selection acts on an entire organism and not its individual parts (Luria et al. 1981; Fish and Beneski 2014). Biological organisms are a mosaic of integrated components that function to achieve evolutionary success (i.e., survive and reproduce). Some of these components may be at odds with other features of an organism. Therefore, organisms must compromise optimal solutions for the necessity of having an integrated system that can perform a number of simultaneous functions (Katz and Jordan 1997; Webb 1997; Fish and Beneski 2014). Thus, evolution rarely leads to solutions with a maximal performance and with an economy of resources. Gould (1983) remarked that organisms are not optimal machines and do not approach engineering perfection.

Because natural selection selects from a diversity of design and performance possibilities controlled by the genetic code and functional demand of the local environment, a multiplicity of possible solutions to engineering problems are potentially available (Fish 2000, 2006). However as with technology, the laws of physics and the physical properties of environment and structural materials available to biological forms impose constraints on evolution (Alexander 1985). All possible structures and processes that potentially could benefit an organism are not available. In addition, radical redesigns are not permitted to expedite enhancing performance as evolution modifies small changes in the genetic code. Existing designs are modified in accordance with the evolutionary pathway along which organisms have evolved (Vogel

1998; Fish 2000; Fish and Beneski 2014). As evolution is descent with modification, the evolutionary history will constrain design. Locomotion by whales would be more efficient if these animals avoided the surface of the water and remained submerged like fish, because of increased wave drag when swimming close to the water surface (Hertel 1966). However, whales as obligate air-breathing mammals and share a common evolutionary history with other tetrapods that have lungs for respiration. Whales must periodically return to the water surface to fill their lungs despite the increased energy cost of this action (Berta 2012).

With such limitations on the products of evolutionary innovation, copying nature for engineering purposes to date has rarely produced practical results (Vogel 1998). A notable exception is the engineering application based on the design of the flippers of the humpback whale with the presence of rounded bumps, called tubercles, along the LE (Miklosovic et al. 2004; Fish et al. 2011a, b). Tubercles represent a highly novel, passive means of altering fluid flow around a wing-like structure that can delay stall and simultaneously increase lift and reduce drag (Fish et al. 2011a, b). With these hydrodynamic advantages, the tubercles are important in the biology of the humpback whale and have biomimetic engineering applications.

3　Biology

3.1　Humpback Whale

The humpback whale (*Megaptera novaeangliae*) is a member of the aquatic mammalian order Cetacea, which includes whales, dolphins and porpoises. Within this grouping, humpback whales are related to the baleen whales of the Mysticete and specifically the rorquals of the taxonomic family Balaenopteridae. This family includes the blue whale (*Balaenoptera musculus*), finback (*Balaenoptera physalus*), and minke (*Balaenoptera acutorostrata*) whales. As with all cetaceans, the rorquals swim by dorso-ventral oscillations of the wing-like tail flukes and use their mobile pectoral flippers as control surfaces for maneuvering (Fish and Rohr 1999; Fish 2002; Cooper et al. 2008; Weber et al. 2009, 2014; Segre et al. 2016; Fish and Lauder 2017). The flippers are homologous to the human arm and hand (manus) with a similar skeletal structure (i.e., humerus, radius and ulna, carpals, metacarpals, and phalanges) (Cooper et al. 2007). The rorquals are large oceanic predators that prey on planktonic crustaceans and small fish. To capture their prey, rorquals lunge rapidly with their mouths open. The force of the water and prey enormously expands the throat region composed of longitudinal pleats (Goldbogen et al. 2006, 2008; Potvin et al. 2009, 2012; Shadwick et al. 2013). With the mouth closed, the water is forced out between the baleen plates leaving the food to be swallowed.

Although not the largest of the rorquals, that being the blue whale, the humpback whale is a relatively short and stouter whale compared to the other rorquals (Winn and Reichley 1985). The humpback whale reaches 19 m in length and exceeds 53 tons in weight. The distribution of the whale extends to all oceans and into the polar regions to the edge of the ice (Minasian et al. 1984). The whales feed in the higher latitudes for half of the year, and then migrate to the tropics for the remainder of the year to mate and give birth. Humpback whales can swim at 2.6 m/s when feeding (Jurasz and Jurasz 1979), 1.1–4.2 m/s when migrating (Chittleborough 1953) and burst at 7.5 m/s (Tomilin 1957).

3.2   Flippers and Tubercles

The flippers of the humpback whale are the longest of any cetacean both in absolute length and relative to body size (Fig. 1; Fish and Battle 1995). Flipper length (i.e., span) can be over 5 m. The length of the flipper varies from 0.25 to 0.33 of total body length (Tomilin 1957; Winn and Winn 1985; Edel and Winn 1978; Fish and Battle 1995). The aspect ratio (flipper span^2/planar area) of the flippers varies from 3.6 to 7.7. This geometry gives the flipper a wing-like appearance that is the basis of its scientific name, *Megaptera novaeangliae*, which means "great wing of New England." Despite their length, the flippers are highly mobile at the shoulder and display some flexibility along their length (Edel and Winn 1978). In cross-section (Fig. 2), the flippers resemble a streamlined fusiform body with a rounded nose or LE increasing to a maximum thickness at 21% of the chord (i.e., linear distance from LE to trailing edge (TE)) and then slowly tapering to a pointed TE. The flipper cross-section is similar to the low drag NACA 63$_4$-021 wing (Abbott and von Doenhoff 1949; Fish and Battle 1995).

Another descriptive name that had been used for the humpback whale was *Megaptera nodosa*. For this case, the species epithet referred to the knobby swellings or bumps that occur on the head and LE of the flippers (Bonner 1989). Whereas typical wing-like structures have a straight LE without the presence of irregularities or perturbations, the humpback flipper defies convention with prominent rounded

Fig. 1 Images of humpback whales' flippers showing LE tubercles. Courtesy of William Rossiter

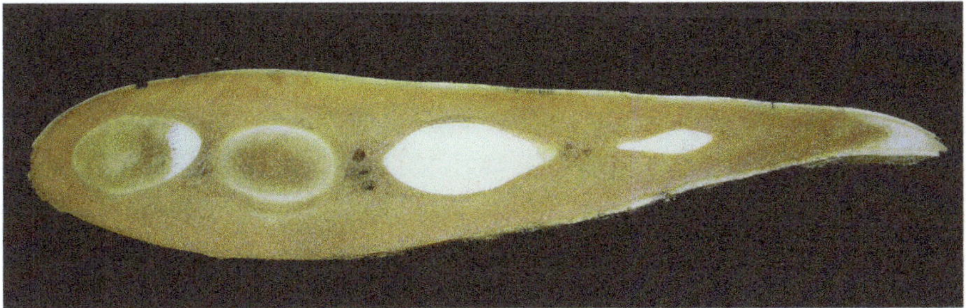

Fig. 2 Cross-section of humpback whale flipper showing streamlined, fusiform profile. The four white areas in the section are bones associated with the phalanges

bumps along the LE (Fig. 1) of its high-aspect ratio (i.e., long and thin) flipper (Fish et al. 2008). There are 9–11 sinusoidally arranged tubercles positioned on the LE of the flipper giving the flippers a scalloped appearance (Winn and Reichley 1985; Bushnell and Moore 1991; Fish and Battle 1995). The geometry of the tubercles presents a series of peaks and troughs. Tubercle development occurs early in utero as the tubercles are displayed in young fetuses (Fig. 3).

The position of the tubercles is related to the occurrence of the joints and terminal phalanges associated with the hyperphalangy of the manus (Cooper et al. 2007). Hyperphalangy is an increased number of phalangeal elements per digit beyond the normal (2/3/3/3/3) condition for mammals. Humpback whales have only four digits in the manus with the absence of digit I (Cooper et al. 2007). Tubercles are large in size near the body but decrease in size toward the tip of the flipper.

The first and fourth tubercles from the shoulder are the most prominent (Fig. 4).

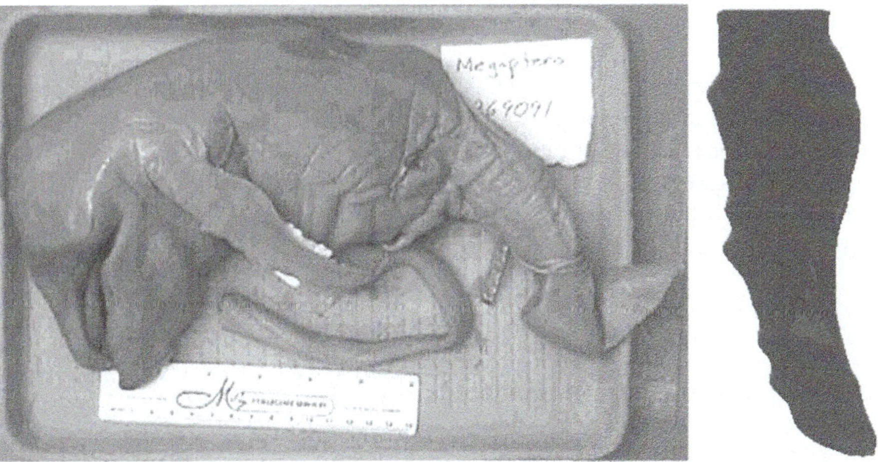

Fig. 3 Fetal humpback whale showing the elongate flipper and tubercles (left). Reconstruction of the fetal flipper from CT scans. Tubercles are prominent and the pattern is the same as in Fig. 1

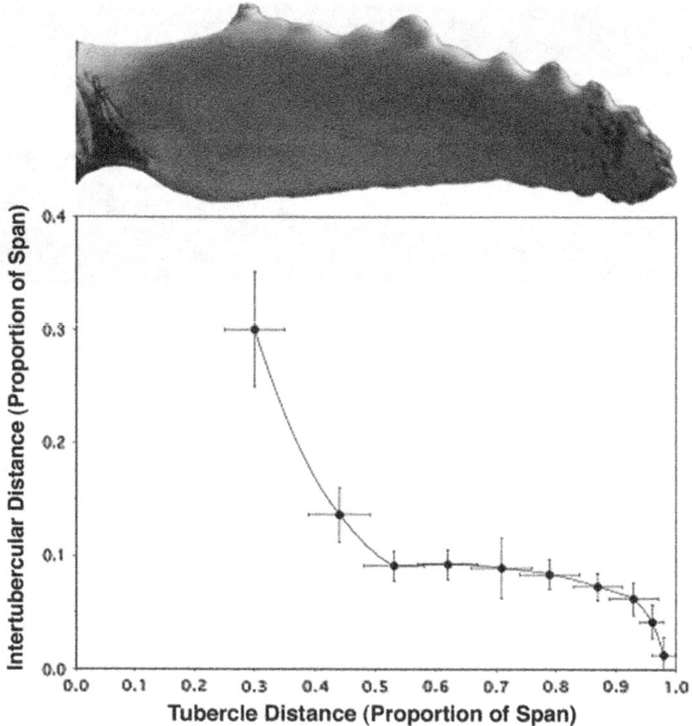

Fig. 4 Relationship of the intertubercular distance with the proportion of flipper span. The error bars are ± one standard deviation of the mean. The flipper planform above the graph illustrates the distribution of tubercles along the LE

The first tubercle occurs at the joint between the radius and metacarpal of digit II. The fourth tubercle is positioned at the terminus of digit II. Extending forward of the LE of the flipper, the first and fourth tubercles are 14.9 and 19.4% of the local chord, respectively. The extension of these tubercles beyond the LE of the flipper is about 4% of flipper span. The other tubercles are smaller with decreasing size toward the flipper tip. The distances between tubercles decrease distally along the flipper. This intertubercular distance is, however, relatively uniform between 6.5 and 8.6% of the span at the mid-span of the flipper (Fig. 4; Fish and Battle 1995; Fish et al. 2011a, b).

3.3 Swimming Performance and Foraging Behavior

The swimming performance of humpback whales shows them to be highly maneuverable and aquabatic, employing their extremely mobile and elongate flippers for banking and turning maneuvers (Tomilin 1957; Nishiwaki 1972; Edel and Winn 1978;

Madsen and Herman 1980; Fish et al. 2008; Friedlaender et al. 2009; Wiley et al. 2011). Humpback whales use their flippers as biological hydroplanes to achieve tight turns associated with their specialized feeding behaviors (Hain et al. 1982; Hazen et al. 2009; Fish et al. 2011a, b).

Generally, balaenopterids (e.g., blue, fin, minke) forage by swimming rapidly forward to engulf prey-laden water (Pivorunas 1979; Ridgway and Harrison 1985; Woodward et al. 2006; Cooper et al. 2008). This typical feeding behavior for rorquals requires that they swim rectilinearly with little maneuvering. Humpback whales are different from the other rorquals in their feeding strategy. Humpback whales rely on tight, rapid turns to capture prey (Fish and Battle 1995). Specialized feeding behaviors that require maneuvering include the "inside loop" and "bubbling" behaviors.

The "inside loop" behavior starts with slapping the water surface with its flukes to startle the prey into a dense aggregate (Hain et al. 1982). The whale then swims rapidly away from the prey aggregate with its flippers abducted and protracted (Edel and Winn 1978). Subsequently, the whale makes a sharp 180° U-turn ("inside loop"), and lunges toward the prey (Hain et al. 1982).

During "bubbling" behaviors, underwater exhalations from the paired blowholes produce bubble clouds or rising bubble columns, which concentrate the prey into a tight ball (Winn and Reichley 1985; Sharpe and Dill 1997; Leighton et al. 2007; Reidenberg and Laitman 2007). The most dramatic bubbling behavior is the creation of bubble nets. These bubble nets concentrate and corral the prey inside a circular pattern of bubble columns. The bubble net is produced as the whale releases bubbles and swims toward the surface in a circular pattern from depth. At completion of the bubble net, the whale pivots using its flippers and banks to the inside of the circular net as it turns sharply into and through the center of the bubble net (Ingebrigtsen 1929; Hain et al. 1982). Groups of whales can work cooperatively when bubble netting. The smallest bubble net recorded had a diameter of 1.5 m and the largest was 50 m (Fig. 5; Jurasz and Jurasz 1979). Based on use of the flippers alone, the minimum calculated turn diameter with a 90° bank angle is 14.8 m (Fish and Rohr 1999).

Turning is important in the capture of elusive prey by the humpback whale. The whale feeds on a variety of foods, including Antarctic krill (euphausiids) and schooling fish, such as sardine, mackerel, anchovy and capelin (Bonner 1989). The elongate flippers function as wings to generate the forces necessary for turning maneuvers (Fish et al. 2011a, b). The sharp, high-speed banking turns that are executed by the humpback whale are enhanced by the high lift/drag characteristic of the high-aspect ratio flippers. In a banking turn, the body rolls on its side and tilts toward the inside of the turn.

During banked turns, the lift force developed by the flippers has a horizontal component directed toward the center of the turn that supplies the centripetal force necessary to maintain the turn (Fig. 5; Weihs 1981, 1993; Fish and Battle 1995). Lift and bank angle are inversely related to turn radius (Alexander 1983; Norberg 1990). In addition, increased angle of attack (i.e., AoA, angle between the chord of the flipper and incident water flow) up to the stall point of the wing increases the lift to produce tighter turns. Stall is the point at which there is dramatic loss of lift as

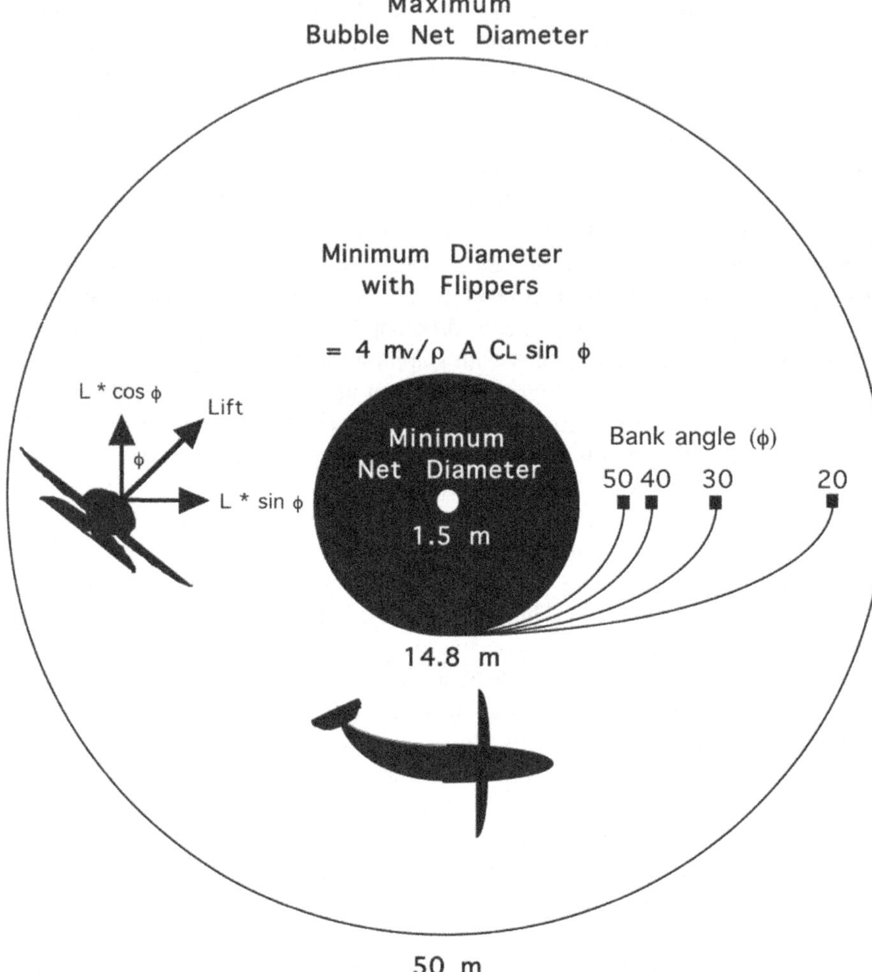

Fig. 5 Calculated and observed turning performance of the humpback whale (*Megaptera novaeangliae*). The calculated minimum turning diameter (14.8 m) for a 9 m whale using only its flippers is shown by the outer margin of the black circle, based on the equation shown in the illustration. The margins of the turn for various bank angles are shown by the curved lines. The minimum and maximum diameters of bubble nets are shown by the margin of the central white circle and the outer white circle, respectively. The lift (L) vectors with respect to bank angle, ϕ, are illustrated in the inset. The turn radius is determined by the equation where m_v is the virtual mass (mass plus added mass of the whale), ρ is the water density, **A** is projected area of the flippers, and C_L is the lift coefficient of the flippers. The silhouette indicates the dimensions of the whale. Figure and equation are from Fish and Rohr (1999)

the AoA is increased. Hydroplanes used in turning must operate at high AoAs while maintaining lift (von Mises 1945; Weihs 1993). If the flippers of the whale were canted at too high an AoA during a turn and stall, the whale would have a reduced centripetal force. This loss of lift and associated centripetal force are analogous to driving a car along a curved road and slipping on a patch of ice. The reduced friction between the ice and the tires would cause the car to drive off the road tangential to the curve, rather than following the original curved trajectory. For the whale, the inability to turn tightly would increase the radius of the turn and potentially allow the prey to escape.

3.4 Hydrodynamic Implications

The prominence of the tubercles on the LE of the humpback whale flippers and the swimming pattern of the whale suggests that these novel structures have a distinct hydrodynamic effect. The knobby leading edge permits the whale to make sharp, high speed banking turns to catch small elusive prey (Fish et al. 2011a, b). Bushnell and Moore (1991) suggested that the knobby LE would reduce drag on the flipper. Later, Fish and Battle (1995) suggested that the ability of the humpback whale to turn tightly was enhanced by the combination of its flippers with the LE tubercles.

An indication that the tubercles had a hydrodynamic effect was based on the distribution of barnacles of the flippers. Balanomorph barnacles were confined to the peak of the tubercles and did not occur between tubercles (Fig. 6; Fish and Battle 1995). As barnacle larvae cannot settle on a substrate if the flow is too high (Crisp 1955), there is an indication that the flow over the flipper is affected by tubercles. Barnacle larvae (cyprids) fail to attach to areas of a water flow exceeding 1– 2 m/s (Crisp 1955; Crisp and Stubbings 1957; Lewis 1978). Typically, when present, barnacles occur on the LE of the tubercles of humpback whale flippers (Edel and Winn 1978; True 1983; Winn and Reichley 1985). Lack of barnacles between tubercles (Fig. 6) was observed by Fish and Battle (1995). This observation indirectly supported the idea a hydrodynamic effect by the tubercles.

3.5 Other Biological Leading-Edge Structures

While the presence of LE tubercles for flow modification is rare, it is not novel. Occurrences of biological LE structures with possible fluid dynamic effects, like the tubercles, has been displayed in a number of animals operating in a fluid environment. During the Paleozoic Era, cartilaginous fishes of the order Iniopterygia exhibited an array of large fish-hook-shaped denticles (i.e., scales) along the LE of their elongate

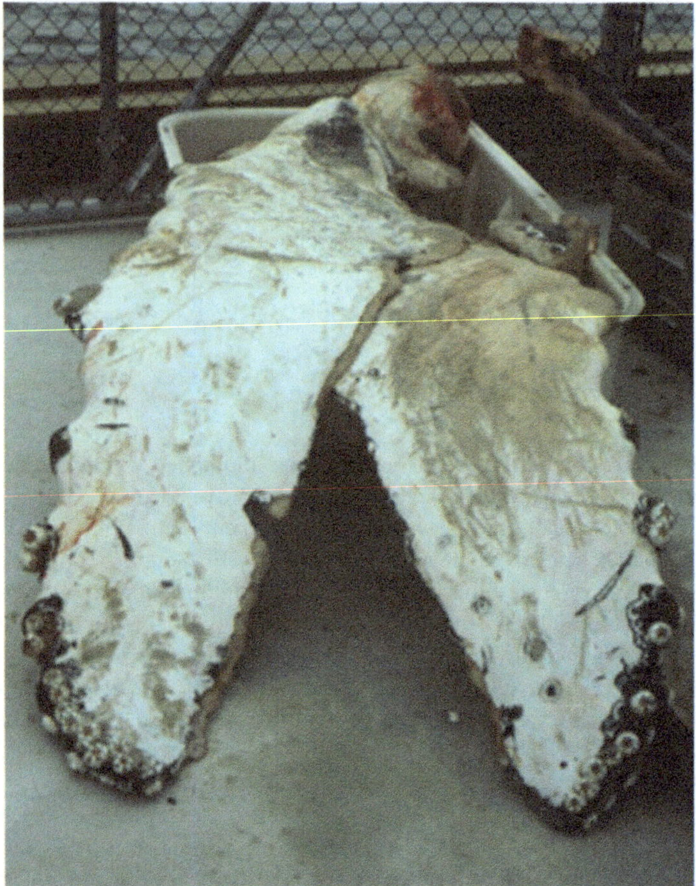

Fig. 6 Image of flippers removed from stranded whale. The barnacles are found on the tubercles except near the flipper tips where the barnacles are more densely packed

pectoral fins (Zangerl and Case 1973). These fishes were envisioned to use the fin as their principle locomotor organ by oscillating the fins vertically in a flapping motion. The genus *Protosphyraena* was a group of swordfish-like, predatory marine fishes from the Upper Cretaceous period. These fishes possessed high aspect ratio pectoral fins with serrated LEs (Fish et al. 2011a, b). Like the humpback whale, the fins of *Protosphyraena* may have been used to maneuver and make tight turns to capture small, elusive prey. The cephalofoil of the hammerhead sharks (e.g., *Sphyrna lewini*) has a scalloped LE (Bushnell and Moore 1991; Kajiura et al. 2003), which may improve hydrodynamic performance related particularly to pitching maneuvers (Nakaya 1995).

In marine mammals other than the humpback whale, LE structures are also found that can modify flow over a hydrodynamic surface. Small tubercles (1.1 mm or less) are present along the LE of the dorsal fin of several porpoise species (Fig. 7;

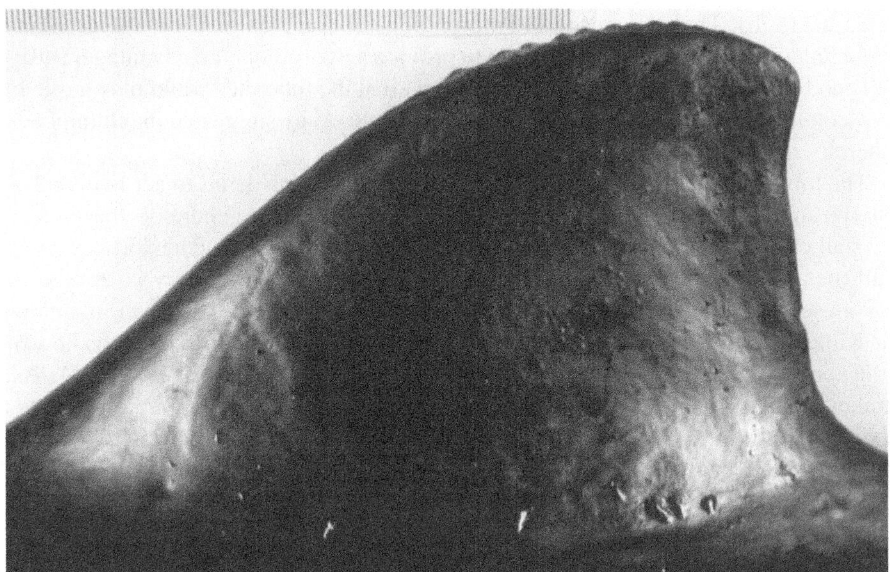

Fig. 7 Dorsal fin of a harbor porpoise showing micro-tubercles at the tip and along the LE

Ginter et al. 2011). It was argued that these microtubercles could act as passive flow-regulating structures to reduce disturbances at the water surface. The mystacial vibrissae of phocid seals have a sinusoidal profile that reduces vortex-induced vibrations to improve sensing of turbulence in the wake of prey fish (Fish et al. 2008; Ginter et al. 2010; Hanke et al. 2010).

The tubercles have been demonstrated to affect drag on biological flyers. The Brazilian (or Mexican) free-tailed bat (*Tadarida brasiliensis*) is a member of the taxonomic family Molossidae. Bats of this family are generally adapted for fast flight (Vaughan 1986). Specifically, the Brazilian free-tailed bat was recorded to fly at a maximum speed 44.5 m/s (McCracken et al. 2018). This bat species has LE tubercles along the dorsal edge of the ears that are similar to the pattern exhibited by humpback whales (Petrin et al. 2018). The tubercles on the bat ear was shown to reduce drag at low AoAs within the operational range of Reynolds number relative to an ear without tubercles (Petrin et al. 2018).

4 Engineering

4.1 Early Studies

The number and position of tubercles on the LE of a humpback whale flipper suggest analogues with specialized LE control devices associated with improvements in hydrodynamic performance (Bushnell and Moore 1991; Fish and Battle 1995; Fish et al. 2011a, b). The occurrence of "morphological complexities" on a hydroplane could reduce, or use, pressure variation to provide a hydrodynamic advantage. Bushnell and Moore (1991) were the first to suggest that the tubercles could play a role in flow control over the flipper of the humpback whale. They suggested that humpback tubercles might reduce drag due to lift on the flipper.

The tubercles of the humpback whale flipper were considered to act like strakes used on aircraft (Fish and Battle 1995). Strakes are large vortex generators that change the stall characteristics of a wing (Hoerner 1965; Shevell 1986). The vortices delay stall (i.e., loss of lift) by exchanging momentum within the boundary layer to keep it from separating from the wing surface. A fully attached boundary layer maintains the Kutta condition so that lift is maintained. The Kutta condition refers to the flow of fluid about a wing that smoothly leaves from the TE (Wegener 1991). At high AoAs, strakes delay stall compared to wings without strakes, although maximum lift does not increase from flow modifications by strakes (Hoerner 1965; Shevell 1986).

The hydrodynamics effect of tubercles was indicated by flow visualization experiments on wavy bluff bodies (Owen et al. 2000). A cylinder that was bent into a spanwise sinusoidal form showed periodic differences in the wake width along the span. A wide wake with two simultaneous vortices occurred where the body protruded downstream and a narrow wake occurred where the body protruded upstream. The vortex represents a region of circular irrotational flow around a central axis (Vogel 1994). The flow in the wake of the wavy body was different from the wake of a straight cylinder, which exhibited a typical von Karman Vortex Street of alternating vortex pairs in the wake. A 30% drag reduction was achieved on a bluff body with a wavy LE compared to an equivalent straight body (Bearman and Owens 1998).

Watts and Fish (2001) conducted the first study related to the unique flipper design of the humpback whale using a computational fluid dynamic (CFD) model. An inviscid 3D panel method was used to model the flow field over wing sections with and without tubercles at an AoA of $\alpha = 10°$. It was found that the trough between tubercles experienced 10% higher total shear stress than other locations along the LE (Fig. 8). Results of the analysis demonstrated that a wing with LE protuberances would increase lift while reducing induced drag compared to a wing with a smooth LE. The panel method showed a 4.8% increase in lift, a 10.9% reduction in induced drag, and a 17.6% increase in lift to drag ratio for wing sections with tubercles (Watts and Fish 2001). It was also speculated that the tubercles could alter, delay, or reduce boundary layer separation and thus delay stall.

A general-purpose incompressible unsteady Reynolds-Averaged Navier-Strokes (RANS) simulation (CFDSHIP; Paterson et al. 2003) was undertaken by John Reifenberg of Carnegie Mellon University and Eric Paterson at Pennsylvania State University (unpubl., 2002). Their analysis was employed to examine the effects of tubercles on flow separation and hydrofoil performance for a symmetrical NACA 63-021 baseline foil at a 10° AoA. The simulated flow conditions were set to a Reynolds number (Re) of 1,000,000 ($Re = UC/\nu$, where U is the flow velocity, C is the foil chord

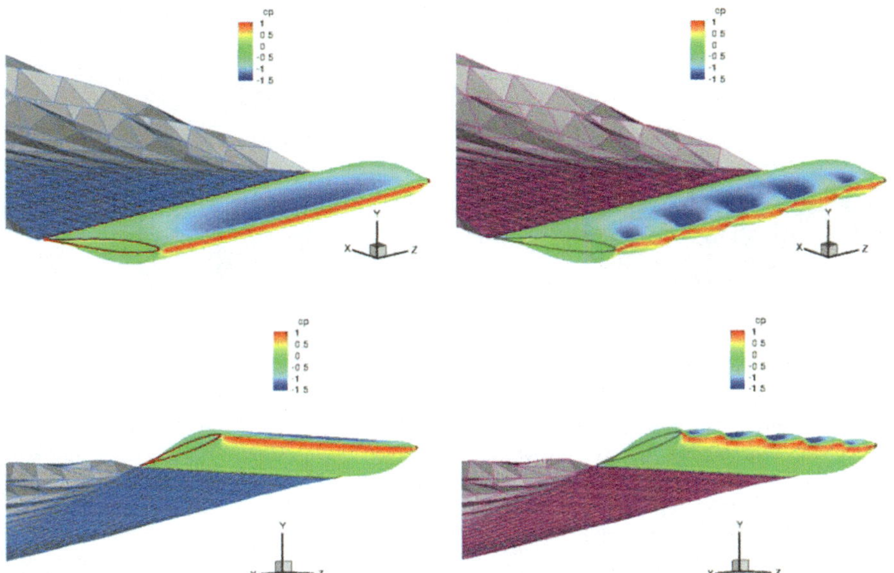

Fig. 8 Panel method simulation of flow over a finite span wing at $\alpha = 10°$ with straight LE (left) and LE tubercles (right) with top (upper) and bottom (lower) views. Colors represent the pressure differences on the wing with red as high pressure, blue as low pressure, and green as intermediate. From Watts and Fish (2001)

length, and ν is the kinematic viscosity of water). Comparisons of wing sections with and without tubercles displayed differing flow patterns that were shown to affect hydrodynamic performance (Fish and Lauder 2006).

The RANS simulation showed that the foil with tubercles had altered surface pressure contours, streamlines, and flow separation pattern in comparison with a foil section with a straight LE (Fig. 9). The straight LE foil showed separation and turbulence anterior of the TE along the span of the foil. For the foil section with tubercles, separation was delayed nearly to the TE for the regions downstream of tubercle crest. This alteration in flow was related to an increase in pressure on the suction side, which locally reduced the adverse pressure gradient.

The tubercles generate separated, chordwise vortices in the troughs at high AoAs (Fig. 9). As the flow does not strike the LE normally, the flow is sheared into the trough's center. As the flow strikes the LE of the trough, two counter-rotating vortices are formed. These vortices are convected along the chord. The spanwise arrangement of the vortices is as a pair on each side of the tubercle crest with opposite spins. The flow over the tubercle crest was juxtaposed between the vortex pair lateral to the flow directly over the tubercle. The tangential velocities of the inward facing flows of the pair of vortices are directed toward the TE of the wing section. The flow from the tubercle peak is accelerated posteriorly due to the interaction with the vortex pair.

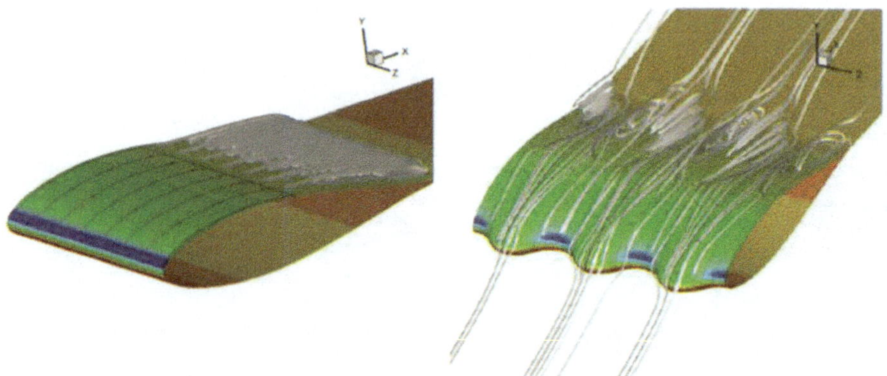

Fig. 9 Pressure contours and streamlines at $\alpha = 10°$ for NACA 63-021 with straight LE (left) and with tubercles (right). An unsteady Reynolds-averaged Navier-Stokes (RANS) simulation was used. A separation line is shown on the wing section without tubercles. For the wing with tubercles, large vortices are formed posterior of the troughs along the LE and flow posterior of the tubercles is shown as straight streamlines without separation. Images courtesy of E. Paterson

These effects prevent the local boundary layer downstream of the tubercles from separating and push the stall line posteriorly toward the TE. When integrated over the entire wing, the structure with tubercles is not prone to stall at higher AoAs than a wing without tubercles (Fish and Lauder 2006).

An empirical test of the tubercle effect was performed using an idealized humpback whale flipper model in a wind tunnel at the United States Naval Academy (Fig. 10; Miklosovic et al. 2004). Two ¼-scale models of a flipper were constructed based on a NACA 0020 section with (scalloped) and without tubercles (baseline). Similar to the real humpback whale flipper, the sinusoidal pattern on the scalloped flipper had an intertubercular spacing and size that decreased with increasing distal location. Static tests were run over a range of AoAs at $Re = 500,000$, which corresponded to speed that was approximately 1/2 of the speed of a whale lunge feeding speed (2.6 m/s).

Wind tunnel tests indicated that wings with tubercles improved maximum lift, increased the stall angle, and kept drag low post-stall (Fig. 10; Miklosovic et al. 2004). The lift coefficient (C_L) increased linearly with increasing AoA for both scalloped and baseline flippers (Miklosovic et al. 2004) up to the point of stall onset. There was a 6% increase in the maximum lift for the scalloped flipper over baseline. The baseline flipper stalled abruptly at an AoA of 11° (Fig. 10). However, the stall angle was increased by 40% for the scalloped flipper with LE tubercles compared to the baseline flipper. When the scalloped flipper did stall, it occurred gradually. The drag coefficient (C_D) of the scalloped flipper was the same as the baseline flipper up until the baseline flipper stalled. This trend indicated that there was no drag penalty for having the tubercles at low α. C_D for the scalloped flipper was less by as much as 32% than that of the baseline geometry in the range $12° < \alpha < 17°$. The lift to drag

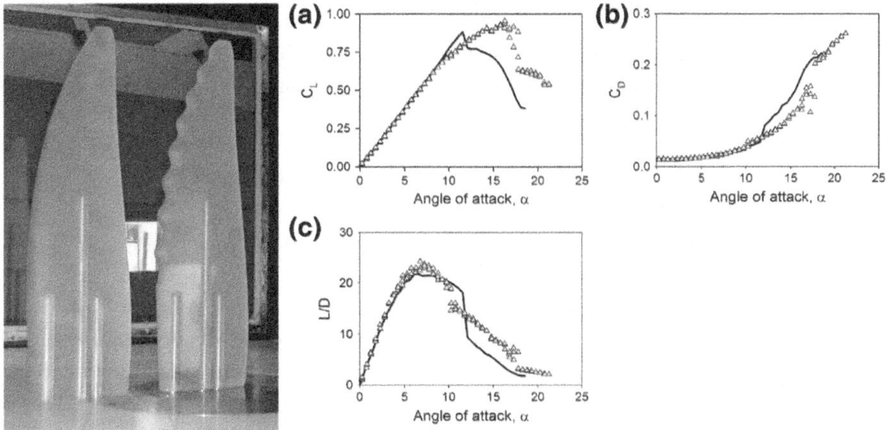

Fig. 10 Lift and drag data for humpback whale flipper models, with and without tubercles, tested in a wind tunnel. The solid lines in A, B, and C show the averages of the data for the flipper model without tubercles and the symbols represent values for the model with tubercles. **a** Wind tunnel measurements of lift coefficient, C_L, displayed as a function of AoA, α. The model with the tubercles maintains lift to higher α than the model without tubercles. The effect of the tubercles is to delay stall. **b** Drag coefficient, C_D, shows no difference between the models up to $\alpha = 11°$. At higher α, the model with tubercles has a lower C_D than the model without tubercles. **c** Aerodynamic efficiency, L/D. **d** Model with and without tubercles. From Miklosovic et al. (2004)

(L/D) ratio, which quantifies the aerodynamic efficiency, displayed a greater peak for the scalloped geometry (Fig. 10; Miklosovic et al. 2004; Hansen et al. 2009).

4.2 Wind and Water Tunnel Testing

Subsequent to 2004, additional studies validated the supposition that the tubercles delay stall for wing-like structures. Increased lift and reduced drag due to the tubercles were more problematic. Lift and drag depended on the experimental flow conditions, geometry of the tubercles, and design of the wing or foil section being tested.

The size and frequency of the tubercles along the LE influences the performance of a wing. The effect of LE tubercles with differing geometries was investigated on the performance of two-dimensional foils based on NACA 0021, 63_4-021 and 65-021 sections (Johari et al. 2007; Custodio et al. 2010; Hansen et al. 2009, 2011). Water tunnel experiments showed that foils with LE tubercles did not stall like foils with a straight LE. The model wing section with LE tubercles at low speeds showed flow separation from the troughs between adjacent tubercles, but attached flow on the tubercles (Johari et al. 2007). The best performance was observed for small amplitude tubercles with regard to lift and stall characteristics (Johari et al. 2007; Hansen et al.

2009, 2011). The wavelength and thus frequency of tubercles was found, however, to have little effect on performance (Johari et al. 2007). Stall was delayed to higher α with C_L over 50% higher for foils with tubercles compared to the baseline foil, but with greater C_D in the prestall regime (Johari et al. 2007). The increased drag was assumed to be due to the lack of a wing tip on the experimental foil sections. The foil tested by Johari et al. (2007) and Custodio et al. (2010) was bounded by an end plate and the wall of the water tunnel.

Fully three-dimensional wings with unmodified wing tips show increased induced drag. The induced drag is associated with the creation of a wing tip vortex. The wing tip vortex is produced from the pressure differential between the upper and lower wing surfaces that induces a rotational flow around the wing tip that is convected downstream (Vogel 1981). The flow pattern set up by the tubercles compartmentalizes the chordwise flow to reduce the tip vortex and corresponding induced drag (Ozen and Rockwell 2010; Fish et al. 2011a, b; Bolzon et al. 2015). The vortices produced between the tubercles along the wingspan effectively act as aero/hydrodynamic fences (Pedro and Kobayashi 2008; Fish et al. 2011a, b; Lohry et al. 2013; Bolzon et al. 2015). Fences act to make the flow over a wing more two-dimensional by maintaining chordwise flow and reducing spanwise flow and thus reduce the wing tip vortex. As the vortices from the tubercles are fluid and not a physical barrier, which increases frictional drag, the vortices act as a virtual barrier and regulate flow. In addition, the flow is compartmentalized to prevent any area along the span that stalls to not affect the rest of the wing (Bolzon et al. 2015).

The tubercle effect is further enhanced with sweepback for a wing (Murray et al. 2005). Model wings with tubercles and sweep angles of 15° and 30° required higher AoAs to achieve stall. Furthermore, the scalloped models showed superior drag performance over models without tubercles (Murray et al. 2005). Flow tests on delta wings with a sweep of 50° demonstrated that at high AoAs large-scale, three-dimensional separation occurred for the wing with a straight LE (Goruney and Rockwell 2009). However, with the addition of tubercles, the flow was radically transformed. Tubercles with an amplitude of 4% of wing chord could completely eradicate the negative effects of the separation and foster re-attachment. The typical thick boundary layer that develops near the wing tip was diminished by the tubercles (Bolzon et al. 2015). The thick boundary layer was responsible for premature stall on a straight-edged sweep wing. Attached flow is maintained at higher AoAs with tubercles in concert with a swept wing to reduce drag and increase lift (Bolzon et al. 2015).

Experiments performed on flapping wings with tubercles showed an effect on the spanwise flow (Ozen and Rockwell 2010). Spanwise flow is a key feature along flapping wings with a straight LE (Ozen and Rockwell 2010). Typically a straight wing, whether flapping or static, will develop a spanwise flow due to the pressure differential that develops between the suction and pressure wing surfaces. Spanwise flow reduces the efficiency of a wing. A flapping wing with tubercles, however, does not produce a pronounced region of spanwise flow as flow is directed in a more chordwise direction. Ozen and Rockwell (2010) also found that the structure of the tip vortex was relatively uninfluenced by the geometry of the LE. The pressure field induced by the flapping motion may have overwhelmed any effect of the tubercles in elevating formation of the tip vortex.

Force measurements demonstrated that the tubercles did not improve hydrodynamic performance in experiments performed on foils with tubercles oscillating in roll and pitch (Stanway 2008). Stanway (2008) considered that degraded performance in flapping was from the redirection of energy to tubercle-generated vortices from the vortices of the wake, which are necessary for thrust production during flapping. Alternatively, the limitation of the tubercles in flapping was assumed be due to the period of oscillations being too rapid to allow the full development of the vortices over a wing with their benefits as has been observed from static wing tests (Fish et al. 2011b). It is necessary to have a relatively steady flow to maintain the pattern of the vortices and incur the hydrodynamic advantages (Stanway 2008).

4.3 Computational Fluid Dynamics

The use of computational methods has been beneficial in understanding the fluid mechanics and modifying the complex geometries associated with the humpback whale tubercles (van Nierop et al. 2008; Pedro and Kobayashi 2008; Saadat et al. 2010). Various computational methods (e.g., Panel Method, RANS, LES, Detached Eddy Simulation) have validated the empirical experiments and arrived at similar conclusions regarding the aero/hydrodynamic performance of the tubercles and the flow patterns produced (van Nierop et al. 2008; Pedro and Kobayashi 2008; Ozen and Rockwell 2010; Saadat et al. 2010; Dropkin et al. 2012; Lohry et al. 2012; Zhang et al. 2012; Câmara and Sousa 2013; Corsini et al. 2013; Xingwei et al. 2013; Serson et al. 2017; Filho et al. 2018). In some cases, however, the computational studies did not fully emulate the wind/water tunnel studies of full three-dimensional wings as these simulations were for infinite wings (Watts and Fish 2001; Bolzon et al. 2015).

When a computational model (Detached Eddy Simulation) used the same geometry that was previously tested in wind tunnel (Miklosovic et al. 2004), the results were in good agreement (Pedro and Kobayashi 2008). The vortices produced from the tubercles were considered to re-energize the boundary layer by carrying high momentum flow close to the flipper's surface (Pedro and Kobayashi 2008; Hansen et al 2010; Bolzon et al. 2015). In addition, the aerodynamics is improved by confining separation to the tip region. Using geometry replicated previously in a wind tunnel test (Miklosovic et al. 2004), another computational model (Unsteady RANS using the k-ω and Spalart-Allmaras turbulence models) also found that tubercles delayed stall by causing a greater portion of the flow to remain attached on a flipper with tubercles as compared to a flipper without tubercles (Weber et al. 2011). This work also found that the attached flow was localized behind the tubercle crests.

The mechanism for enhanced hydrodynamic performance due to the presence of tubercles appears due to the specific pattern of vortex generation over the surface of the flipper. Computational studies indicated that the vortices produced from the tubercles re-energize the boundary layer by carrying high momentum flow close to the flipper's surface (Wu et al. 1991; Pedro and Kobayashi 2008; Hansen et al. 2010). This new flow pattern is determined by the geometry of the tubercled LE, particularly with respect to the flow in the troughs between the peaks of adjacent tubercles. As the flow impacts the LE in the troughs between two tubercles, it is deflected into the center of the trough producing a pair of organized vortices with opposite spins (Hansen et al. 2010; Fish et al. 2011a, b). The vortices represent a sacrificed separation. The tangential velocities of the inward facing flows of a pair of vortices from two troughs are directed toward the TE of the wing section. Sandwiched between the two vortices, the flow over the tubercle is energized and accelerated. These effects prevent the local boundary layer from separating and push the stall line further posteriorly on the flipper. This action is analogous to a baseball-pitching machine, which accelerates a ball by squeezing it between two counter-rotating wheels (Fish et al. 2011a). Tubercles delay stall by causing a greater portion of the flow to remain attached on a wing, with the attached flow localized behind the tubercle peaks. When integrated over the entire structure, the flipper with tubercles will not stall at a higher AoA than a flipper without tubercles.

In addition, the flow dynamics are improved by confining separation to the tip region. Tubercles delay stall by causing a greater portion of the flow to remain attached on a wing, with the attached flow localized behind the tubercle crests (Weber et al. 2011). Numerical simulation indicated that the vortices generated by the tubercles alter the velocity profile of the boundary layer (Rostamzadeh et al. 2014). The changes in velocity magnitude and direction would mix higher momentum fluid into the near wall flow and delay stall (Bolzon et al. 2015).

van Nierop et al. (2008) used a potential flow model to demonstrate that the tubercles delay stall by altering the pressure distribution on a wing. The model showed that the adverse pressure gradient was larger and closer to the LE of the wing behind the troughs than downstream of the tubercles. The absolute pressure is lowest in the troughs. This pressure difference indicated that the flow would separate earlier at a lower AoA behind the troughs than behind the tubercle. The model showed differences due to tubercle design compared to data derived from wind tunnel studies by Johari et al. (2007). The computational model indicated that the wavelength of the tubercles had very little effect, whereas the empirical experiment showed a larger effect on performance due to tubercle wavelength.

5 Applications

The inherent problem in combining biology and engineering to produce biomimetic products is the difference in scale operating for each respective system. Engineered systems are generally large in size and fast in speed compared to biological systems, which are relatively small and slow. This problem of scale for use in biomimicry may be reduced by finding areas of overlap in size and performance between a biological structure and an engineered application.

The presence of LE tubercles on a wing-like structure can have a positive influence on the hydrodynamic performance, despite running counter to traditional views on the performance of wing-like structures. The wings of aircraft, blades of fans, turbines and propellers, and other wing-like structures typically have straight LEs. However, the improvement of the aero/hydrodynamic performance of these structures has potential benefits. The benefits of wing-like structures with the addition of tubercles is increased lift, reduced drag post-stall, separation control, and delay in stall in both air and water. Bolzon et al. (2015) and Aftab et al. (2016) also considered that the flow pattern set up by the tubercles could beneficially affect the pressure drag, skin friction drag, induced drag, heat transfer, dynamic stall, laminar separation bubble, and noise reduction. There are limited number of passive means of altering fluid flow around a wing-like structure. As a result, the application of LE tubercles for passive flow control has potential in the design of aviation and marine technologies, fans, and turbines.

5.1 Scale and the Whale

The size of the humpback whale and its flippers operate near or at the same scale as some engineered devices. The flow experienced by and modified by the tubercles is within the same Reynolds regime that coincides with a large array of engineered applications. The Reynolds number of a flipper is 1.6×10^6 when the whale is lunge feeding (2.6 m/s). Thus, the flipper and tubercles are operating in a turbulent flow regime, which is the standard operating condition for many engineered systems. Furthermore, the tubercle function passively to modify flow and maintain favorable hydrodynamic conditions. Therefore, control systems can be simplified.

5.2 Patents

In 2002, a patent was issued on the application of a scalloped wing LE (Watts and Fish 2002). The patent described the tubercles on a wing as "a plurality of protrusions spaced laterally along the leading edge, the protrusions creating a smoothly varying, alternately forward and rearward sweep (greater and lesser sweep) along the leading edge (relative to the upstream flow direction along the leading edge)." The patent claimed an advantage of increased lift over drag ratios compared to straight wings. In addition, the patent assured that flow created by the LE protrusions limited the creation of line of high static pressure along the LE of a wing-like structure. In addition, it was stated that streamwise vortices were created that would reduce vortex strength of the tip vortex and subsequently decrease induced drag. The invention described by the patent claimed to improve the efficiency of a wing and could be applied to aircraft, watercraft, and land vehicles. Suggested applications included rudders, submarine dive planes and conning towers, sailboat keels and masts, spoilers, stators, rotors, fans (Watts and Fish 2002).

A modification of the original patent was issued in 2013 as a "Turbine and compressor employing tubercle leading edge rotor design" (Dewar et al. 2013). The application was directed to improving the efficiency of wind and water turbines and compressors. The additional lift generated by rotor blades would allow more power to be captured from the available fluid flow.

A patent application was also made for modification of the spokes of a bicycle wheel (Zibkoff 2009). The tubercles were to be placed on thick spokes as a method to reduce drag on the wheel.

5.3 Control Surfaces-Rudders and Dive Planes

The ubiquity of wing-like structures for stability and maneuverability presents an opportunity to enlist tubercles to improve performance. Included in such structures are fixed surfaces, such as stabilizers, spoilers, keels, fins, and skegs, and mobile control surfaces, such as rudders and dive planes. Furthermore, prevention of stall by the tubercle effect would reduce flow-induced vibrations in these structures.

The designers at Feadship De Voogt developed *Breathe*, a concept superyacht that incorporates biomimicry into its design (www.feadship.nl). The stabilizers and steering fins were based on the humpback whale flipper with tubercles (Fish et al. 2011b). Like the humpback whale, the use of tubercles on control surfaces have implications for increased maneuverability. Delay of stall by tubercles on both fixed and mobile control surfaces provides a benefit in tight turning situations. To produce the required centripetal force to effect a turn, a lift force that is directed toward the center of the turn is generated by the control surface. The magnitude of the lift is directly dependent on the AoA with a higher angle providing a greater centripetal force. As stall can be delayed with tubercles, the control surface can operate at higher AoAs producing a tighter turn radius with more control.

The ability to maintain lift through a small radius turn is important in the operation of rudders. A low-aspect ratio rudder with tubercles and an unswept LE generated more lift at AoAs above 22° compared to a smooth rudder at a Reynolds number of 200,000 (Weber et al. 2010). At higher Reynolds numbers, this effect diminishes and the tubercles accelerate the onset of cavitation. A tubercled rudder mounted on a sailboat demonstrated enhanced control with better "grip" in addition to a reduction in drag (Whitehouse 2011). A human-powered submarine, *Umpty Squash*, used tubercled dive planes and rudders (Fig. 11; Fish et al. 2011b). The students of the Sussex County Technical High School constructed the submarine. In 2005, the submarine competed in the International Submarine Races held at the David Taylor Model Basin in Bethesda, Maryland. In this competition, the 3.6 m long submarine was able to able to accomplish a starboard 90° turn with a radius of approximately 7.6 m (2.1 lengths) (Fish et al. 2011b).

Fig. 11 Human-powered submarine (left) with rudder and dive planes with tubercles (right). Courtesy of Chris Land and the Sussex County Technical School

Application of tubercles to the mast of a sailboat could be useful during close reach maneuvers (Fish et al. 2011b). A close reach would have the sail set with a high AoA to the apparent wind. The lack of a thick cross-section by the sail itself may preclude any advantage in lift and stall. However, the presence of tubercles on the mast, representing a bluff body, could have advantages in terms of drag. Bluff bodies, like cylinders, can experience lowered drag when the LE has a sinusoidal design (Bearman and Owens 1998).

Commercially, the company Fluid Earth markets a surfboard skeg (fin) with LE tubercles (Fig. 12; Fish et al. 2011a, b). The addition of tubercles would provide enhanced control during a cutback maneuver. Surfers use a cutback to turn rapidly back into a wave. The surfer uses momentum from his/her downward motion on the face of the wave to climb the wave. Near the wave crest, the surfer rapidly turns again to ride down the slope of the wave. As the maneuver requires a rapid small radius turn, a skeg with tubercles could help to avoid stalling the skeg and maintain the trajectory of the turn.

5.4 Wings

The design of the elongate, high aspect ratio flippers of humpback whales that operate at high Reynolds numbers is directly applicable to aircraft wings. Application of tubercles to delay stall is particularly appropriate for wings operating at high AoAs. As tubercles delay stall at high AoAs, their use on conventional aircraft (Fig. 13) may allow for the replacement of boundary layer control structures, such as flaps and slots. These structures are necessary to prevent stall, particularly at periods of high AoAs such as takeoffs and landings. The elimination of flaps and slots with their associated machinery could reduce the weight of the aircraft and increase fuel economy. Bolzon et al. (2015) considered that interest for aircraft application would be confined to low AoAs, although even a small improvement in aerodynamic efficiency would have a

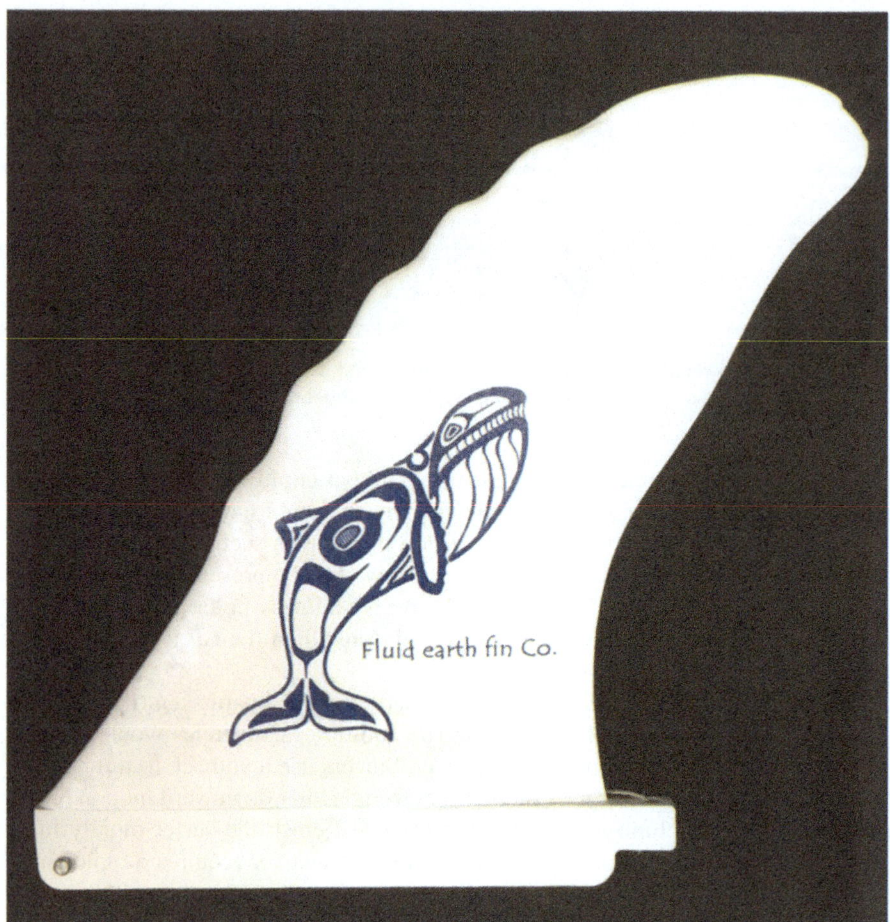

Fig. 12 Surfboard skeg with LE tubercles

substantial economic savings in the commercial airline industry. A patent application (US 8789793 B2; Sandin 2014) was made to apply the LE geometry inspired by the humpback flipper tubercles to an aircraft tail surface. The tubercles would be the site of ice accretion on the tail span. The localized accumulation of ice on the LE would allow channeled airflow and the formation of the vortices to delay boundary layer separation and stall. These effects could make air travel safer by extending the range of AoAs without stall. Furthermore, the presence of tubercles on wings would be advantageous in unsteady and turbulent weather conditions.

Swept wings have earlier stall toward the wing tip, where there is greater loading (Craig 2002; Bertin and Smith 1998). The increased outboard loading promotes premature boundary layer separation, which is amplified by spanwise flow (Bertin

Fig. 13 Model of commercial jet airliner with LE tubercles on the wings and stabilizers. The addition of tubercles on the wings could potentially improve safety, reduce weight, and decrease fuel costs by removal of control surfaces needed to change the stall characteristics of the wings

and Smith 1998). At large AoAs, premature boundary layer separation at the wing tip leads to tip stall. Tip stall can affect the stability of an aircraft wing and promote a loss of effectiveness of control surfaces (Bertin and Smith 1998). To limit spanwise flow, physical boundary layer fences are used on swept wings, however, such fences will increase the frictional drag on the wing.

Murray et al. (2005) investigated the effect of tubercles on wings with sweep of 15° and 30°. Higher AoAs were required by model wings with sweep to achieve stall compared to a models with a straight LE. Specifically, stall was delayed by tubercled models by 40%, 48% and 21% and maximum lift increased by 6%, 9% and 4% for sweep angles of 0°, 15° and 30°, respectively. In addition, swept wings with tubercles demonstrated superior drag performance over the range of AoAs compared to wings without tubercles.

Flow tests on delta wings with a sweep of 50° showed large-scale, three-dimensional separation occurred for a wing with a straight LE at high AoAs (Goruney and Rockwell 2009). With a sinusoidal LE, the flow over a delta wing could be changed and possibly produce more lift. Chen et al. (2013) found that the stall of a delta wing with low sweep (non-slender) could be delayed by a sinusoidal LE with no penalty to the lift produced while lift/drag remained unchanged. The delay in stall was attributed to a difference in the flow pattern between delta wings with a straight-edged and sinusoidal LE (Chen and Wang 2014). At a large AoA, a straight-edged delta wing has a flow that is dominated by large-scale three-dimensional dual LE vortices. With tubercles, it was shown that there were multiple LE vortices on the leeward side of the wing that could explain the delayed stall.

Micro aerial vehicles (MAV) and small, unmanned aerial vehicles (UAV) were tested with tubercles on the LE of the wings (Bolzon et al. 2015; Cai et al. 2017). It was considered that the tubercles would make these vehicles more stable by reducing the chance of stall. The vehicles operate at low Reynolds numbers, where severe stall is more likely to occur. In addition, the size of MAVs and small UAVs make them more prone to stall because of atmospheric turbulence.

5.5 Land Vehicles

The design of trucks encounters large drag and potentially high energy costs for operation. As a means to reduce these costs, methods to control the flow and reduce the drag were targeted for application of tubercles. Models of mirrors with and without tubercles mounted on the side of a truck were tested in a wind tunnel with particle-image velocimetry (PIV) over a range of velocities from 9.55 to 15.0 m/s (Bauer et al. 2013). The mirrors with small tubercles demonstrated reduced drag and less turbulence compared to unmodified mirrors and mirrors with large tubercles.

In 2014, tubercles were added to the rear wing of a Formula One racecar. The top flap of the McLaren MP4-29 had tubercles along the LE (Madier 2014). The top flap is adjustable with respect to the AoA. As conventional wings can only be optimized for maximum efficiency at a narrow range of AoAs, the tubercles would provide a larger range of AoAs to maximize aerodynamic efficiency (Bolzon et al. 2015).

The Zipp company makes bicycle wheels that are based on the humpback whale tubercles. The Zipp 454 NSW wheel has a bumpy profile with the bumps at the position where the spokes of the wheel intersect with the rim of the wheel. The modified wheel is reported to be effective in crosswind conditions (Bradley 2016).

5.6 Propellers

Propellers operating in a marine system have the potential to be improved by the addition of tubercles. The effective AoA of a propeller blade can be increased by increasing the blade angle (Larrabee 1980). A higher AoA can produce more lift to derive greater thrust and increase the effective pitch of the propeller. Ibrahim and

New (2015) performed an investigation using high-fidelity CFD on the hydrodynamics of a marine propeller with tubercles. They concluded that the tubercles produced favorable flow effects by varying the pressure and velocity distributions on the propeller blades. This effect produced a 10% increase in thrust, but with a 5% decrease in propeller efficiency.

Tubercles would be advantageous for a heavily loaded wing, operating at a lower than optimum aerodynamic efficiency (Erickson 1995). Such conditions occur with helicopter rotor blades (Erickson 1995; Conlisk 1997). As a blade is rotating forward in the direction of the helicopter flight, the opposite blade is rotating in the opposite direction. The difference in velocity between the blades can cause dynamic stall in the retreating blade with a temporary loss of lift (Coxworth 2012; Aftab et al. 2016). The loss of lift creates turbulence between the blades, which places stress on the rotors and limits the speed of the helicopter. To compensate for the loss of lift by the retreating blade, the AoA is increased above the static stall angle (Mai et al. 2006). Inspired by the humpback whale tubercles, engineers from the German Aerospace Center adhered small (diameter = 6 mm) flat rubber cylinders along the LEs of the span of the rotor blades (Mai et al. 2006; Coxworth 2012). These microtubercles were referred to as Leading-Edge Vortex Generators (LEVoGs). The LEVoGs on rotor blades were tested in a wind tunnel and flown on a helicopter (Mai et al. 2006; Coxworth 2012). The LEVoGs improved aerodynamic performance by increasing overall time-averaged lift, and decreasing pitching-moment and pressure drag (Mai et al. 2006).

Speedo developed a training swim fin called *Nemesis*. The design of the fin was inspired by the tubercles of the humpback whale flipper. The fins have several large tubercles along the outer edge of the fin (Fig. 14).

5.7 *Fans and Compressors*

The addition of tubercles on ventilation fan blades enhances energy. Envira-North Systems Ltd. produces industrial ceiling fans for large buildings (e.g., factories, warehouses, arenas, dairy barns) with the tubercle modification on the LE of each fan blade. A high-volume, low-speed (HVLS) model with a 24-foot diameter of five blades is reported to be 25% more efficient and consumes 20% less electricity to operate than a 10-blade configuration (Anonymous 2010). In addition, the HVLS fan is 20% quieter. Numerical examination of exhaust fans by Corsini et al. (2014), however, showed only a 1% increase in post-stall performance compared to the baseline.

A RANS simulation performed on a compressor cascade demonstrated that a small wavelength tubercle configuration on the compressor blades had a maximum loss reduction of 46% and increased the stall angle (Zheng et al. 2018). The delay of stall by tubercles is considered to be important in the development of axial compressors in gas turbines (Keerthi et al. 2016). Stall in compressor blades can lead to increased structural load and breakage by the blades. Stall delay would reduce the likelihood of

Fig. 14 Nemesis training fins produced by Speedo and inspired by the humpback whale tubercles

catastrophic effects and more effective operation by allowing for increased pressure rises with fewer steps. Using a cascade of airfoils, Keerthi et al. (2016) found an increase in stall angle by 8.6°. The improved performance was highest with small amplitude and wavelength tubercles.

5.8 Turbines

The desire for clean renewal sources of energy has fueled interest in application of the tubercle technology to power generation by turbines (Abate and Mavris 2018). Tubercle modified blades can be effective in power generation for wind turbines and marine tidal turbines in air and water, respectively (Mueller 2008; Murray et al.

2010; Shi et al. 2016, 2017, 2018; Kulkani et al. 2018). The effective AoA of a turbine blade can be increased, thereby producing more lift to capture more flow energy and increase electrical power generation. The use of tubercles can effectively be employed in the generation of power by turbines in either water or air.

Field trials run on a 35 kW, variable pitch wind turbine with retrofitted blades with tubercles by the WhalePower Corporation, which the author is president, at the Wind Energy Institute of Canada on Prince Edward Island (Fig. 15). The modified wind turbine demonstrated increased electrical generation at moderate wind speeds compared to unmodified blades (Fig. 16; Mueller 2008; Wind Energy Institute of Canada 2008; Howle 2009). In another outdoor test, a wind turbine with nine blades was mounted to a car to vary wind speed (1–7.5 m/s) and test the effectiveness of tubercles (Leung 2014). The blades with tubercles were found to produce more power by 16–30% at 2–6.5 m/s. Gupta et al. (2013) found that tubercled blades performed better than straight blades by maintaining power production during stall conditions.

In the field, wind turbines must operate under unsteady, turbulent, and severe wind conditions. A numerical analysis was conducted on a modified NREL phase VI turbine blade using the commercial code *Fluent* (Zhang et al. 2012). Although the addition of tubercles was found to promote boundary layer separation under the experimental flow conditions, the modified blade was considered to have a more robust power output than unmodified blades on a stall regulated turbine for wind speeds from 10 to 20 m/s. Zhang et al. (2012) recommended that the tubercled blades on a pitch control turbine would be useful in complex unsteady wind conditions. In a modification of the tubercled blade, the tubercles were placed on the TE of blades on a wind turbine and stabilized turbine performance by suppressing turbulence in the wake (Ibrahim et al. 2015). It was considered tubercles in this configuration would be recommended in severe wind conditions with unsteady and high wind velocities.

Fig. 15 Wind turbine blades utilizing LE tubercles

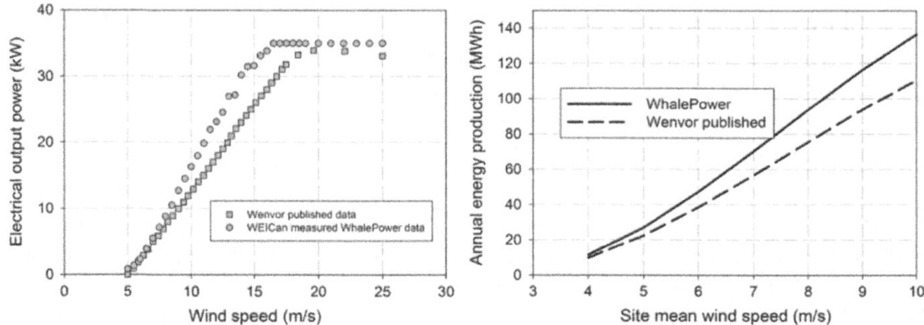

Fig. 16 Comparative performance of wind turbine blades with tubercles versus standard straight blades. The graphs show the results of a study conducted at the Wind Energy Institute of Canada for the WhalePower turbine. The turbine blades were compared between a standard Wenvor blade (squares) and the WhalePower blades with tubercles (circles). With increasing wind speed, the turbine blades using tubercles out-performed the unmodified blades with respect to electrical output power (left) and annual energy production (right) at moderate wind speeds. Courtesy of WhalePower Corporation

Although the humpback whale possesses tubercles over the distal 70% of the flipper (Fish et al. 2011a), this design needs not be explicitly followed for turbine blades. As these blades spin about a central axis, the highest velocities and greatest lift production will occur on the distal-most (outboard) portion of the blade. Zhang and Wu 2012) found that outboard 60% of the blade to the blade tip was the primary region for power generation. Abate and Mavris (2018) showed that by placing tubercles from 95% of the blade span to the tip there was a calculated 10% increase in annual energy production at a wind speed of 10 m/s.

Analogous to wind turbines, marine tidal turbines with tubercle-modified blades were found to generate power at low flow speeds (Murray et al. 2010). Tubercles were placed on the distal 40% of the three turbine blades. Compared to blades with smooth LE, blades with LE tubercles demonstrated enhanced performance.

5.9 Noise Reduction

Prevention of stall, reduction of the tip vortex, and flow modification by the tubercle effect would reduce flow-induced vibrations and noise production. The noise produced by an airfoil is a high-pitched whistle (Hansen et al. 2010). These noises have been associated with wind turbines, gliders, small aircraft, and fans.

Studies by Hansen et al. (2010, 2012) showed a suppression of tonal noise that was possible by the addition of tubercles to a propeller. Effective reduction of tonal noise was found to be reduced by large amplitude and smaller wavelength tubercles.

Using a numerical simulation to study the interaction of tubercles and aerofoil-gust interaction noise, Lau and Kim (2013) found that noise reduction was improved with the presence of the tubercles. Similarly, other computational studies showed that noise reduction occurred due to tubercles modifying the flow field (Kim et al. 2016; Turner and Kim 2017; Wang et al. 2017). Wang et al. (2017) found a decrease in noise by 13.1–13.9 dB with an almost unchanged drag coefficient for a large eddy simulation of a NACA 0012 foil with tubercles and TE serrations. Polacesek et al. (2011) and Clair et al. (2013) studied the tubercle effect on turbo fan blades and found a 3–4 dB reduction of tonal noise without changing aerodynamic performance.

In the marine environment, cavitation is a source of noise for propellers and tidal turbines. The tubercles do not suppress cavitation, but trigger cavitation to particular areas of the blade (Weber et al. 2010; Shi et al. 2016). Because the pressure is lowest in troughs of tubercled wings, cavitation occurs first behind the troughs (van Nierop et al. 2008). As tidal turbines could produce a particular noise signature, noise reduction would be beneficial to marine organisms although the use of radiated noise by whales and seals to avoid collisions with tidal turbines may be negatively impacted. Reduction of noise by propellers would be advantageous for stealth in naval operations and in reducing noise pollution by commercial marine traffic.

6 Conclusion

Insights into the way that animals function are being used in the development and improvement of technology through the biomimetic approach are described here. By imitating nature, alternative solutions to engineering problems can be fostered (Benyus 1997; Vogel 1998; Bar-Cohen 2006; Allen 2010; Fish and Kocak 2011; Fish et al. 2011a). For example, the technology associated with the development of robots is becoming more dependent on biomimetics and biologically inspired designs (Taubes 2000; Gravish and Lauder 2018). As biologists and engineers collaborate, they are developing innovative ideas that have immense application potential. With limited energy reserves and greater competition for this limited resource, greater energy economy and energy production are desired in the construction of engineered systems. Imaginative solutions from nature may serve as the inspiration for new technologies.

The design of the tubercles on the flipper of the humpback whale provides inspiration for the development of a passive means of altering fluid flow around wing-like structures that can delay stall and both increase lift and reduce drag simultaneously. These effects are due to the function of the tubercles as vortex generators. The tubercles are responsible for the formation of paired vortices in the troughs between tubercles. These vortices interact with the flow over the tubercles to keep the flow attached to the wing surface and delay stall. This design is highly novel. The fusion of tubercles with engineered systems has the potential to produce biomimetic designs that exhibit superior performance compared to the current technology.

The aero/hydrodynamic performance by LE tubercles has potential applications for the passive control of flow. Applications of the tubercle technology include fins, rudders, dive planes, control surfaces, wings, propellers, compressors, pumps, wind and water turbines, and noise reduction.

The utility of tubercles in improving the performance of engineered systems was inspired from the examination of an animal that at one time was near annihilation. By the beginning of the 1900s, the population of humpback whales was reduced by whaling to only 10% of its historic numbers. The whales continued to be hunted through most of the 20th century with the other baleen whales. While these whales have roamed the seas for millions of years, it was only recently that the hydrodynamic benefits of tubercles of the humpback whale flipper were recognized. The novel morphology of the humpback whale flipper with tubercles provides a new direction in the design of devices that can help conserve energy as well as cleanly and sustainably generate energy. It is therefore ironic that an animal, which was exploited close to extinction by humans, should provide us with the inspiration to better our own future. The lesson from such cases as the humpback whale is that a diversity of organisms needs to be conserved as a potential source of innovation as the application of biomimetic technology becomes ever more integrated into our lives.

References

Abate G, Mavris DN (2018) Performance analysis of different positions of leading edge tubercles on a wind turbine blade. In: 2018 Wind Energy Symposium, AIAA SciTech Forum, 8–12 January 2018, Kissimmee, FL

Abbott IH, von Doenhoff AE (1949) Theory of wing sections. Dover Publ, New York

Afroz F, Lang A, Habegger ML, Motta P, Hueter (2017) Experimental study of laminar and turbulent boundary layer separation control of shark skin. Bioinsp Biomim 12:016009

Aftab SMA, Razak NA, Mohd Rafie AS, Ahmad KA (2016) Mimicking the humpback whale: an aerodynamic perspective. Prog Aerospace Sci 84:48–69

Alexander RMcN (1983) Animal mechanics. Blackwell, Oxford

Alexander RMcN (1985) The ideal and the feasible: physical constraints on evolution. Biol J Linn Soc 26:345–358

Allen R (2010) Bulletproof feathers. University of Chicago Press, Chicago

Anderson JD Jr (1998) A history of aerodynamics. Cambridge University Press, Cambridge

Anonymous (2010) Energy efficient fans take their cue from the humpback whale. Ontario Power Authority 4:1–4. http://archive.powerauthority.on.ca/Storage/122/16957_AgNews_July231.pdf

Bar-Cohen Y (2006) Biomimetics-using nature to inspire human innovation. Bioinsp Biomim 1:P1–P12

Bar-Cohen Y (2012) Biomimetics: nature-based innovation. CRC, Boca Raton

Bauer D, Dunker T, Grill S, Kuhnel D (2013) Bionische Optimierung von LKW-Seitenspiegeln zur Verringerung des Gesamtwiderstandes (Bionic optimization of truck side mirrors). Report: Hochschule, University of Applied Sciences, Bremen

Bearman PW, Owens JC (1998) Reduction of bluff-body drag and suppression of vortex shedding by the introduction of wavy separation lines. J Fluids Struct 12:123–130

Bechert DW, Bruse M, Hage W (2000) Experiments with three-dimensional riblets as an idealized model of shark skin. Exp Fluids 28:403–412

Benyus J (1997) Biomimicry: innovation inspired by nature. HarperCollins, New York

Berta A (2012) Return to the sea. University of California Press, Berkeley

Bertin JJ, Smith ML (1998) Aerodynamics for engineers. Prentice Hall, Upper Saddle River

Bhushan B (2009) Biomimetics: lessons from nature—an overview. Phil Trans Roy Soc A 367:1445–1486

Bolzon MD, Kelso RM, Arjomandi M (2015) Tubercles and their applications. J Aerosp Eng 29:04015013

Bonner N (1989) Whales of the World. Facts on File, New York

Bradley J (2016) Zipp 454 NSW wheels take design cues from whales. Velonews, http://www.velonews.com/2016/11/news/zipp-454-nsw-wheels-take-design-cues-whales_424186

Burke J (1978) Connections. Little, Brown and Company, Boston

Bushnell DM, Moore KJ (1991) Drag reduction in nature. Ann Rev Fluid Mech 23:65–79

Cai C, Zuo Z, Maeda T, Kamada Y, Li Q, Shimamoto K, Liu S (2017) Periodic and aperiodic flow patterns around an airfoil with leading-edge protuberances. Phys Fluids 29:115110

Câmara JFD, Sousa JMM (2013) Numerical study on the use of a sinusoidal leading edge for passive stall control at low Reynolds number. 51st AIAA Aerospace Sciences Meeting, 7–10 Jan 2013, Grapevine, TX

Chen H, Wang JJ (2014) Vortex structures for flow over a delta wing with sinusoidal leading edge. Exp Fluids 55(6):1–9

Chen H, Pan C, Wang J (2013) Effects of sinusoidal leading edge on delta wing performance and mechanism. Sci China Tech Sci 56(3):772–779

Chittleborough RG (1953) Aerial observations on the humpback whales, *Megaptera nodosa* (Bonnaterre), with notes on other species. Aust J Mar Freshwater Res 4:219–226

Clair V, Polacek C, Le Garrec T, Reboul G, Gruber M, Joseph P (2013) Experimental and numerical investigation of turbulence-airfoil noise reduction using wavy edges. AIAA J. 51(11):2695–2713

Conlisk AT (1997) Modern helicopter aerodynamics. Ann Rev Fluid Mech 29:515–567

Cooper LN, Dawson SD, Reidenberg JS, Berta A (2007) Neuromuscular anatomy and evolution of the cetacean forelimb. Anat Rec 290:1121–1137

Cooper LN, Sedano N, Johannson S, May B, Brown J, Holliday C, Kot W, Fish FE (2008) Hydrodynamic performance of the minke whale (*Balaenoptera acutorostrata*) flipper. J Exp Biol 211(12):1859–1867

Corsini A, Delibra G, Sheard AG (2013) On the role of leading-edge bumps in the control of stall onset in axial fan blades. J Fluids Eng 135(8):081104

Corsini A, Delibra G, Sheard AG (2014) The application of sinusoidal blade-leading edges in a fan-design methodology to improve stall resistance. Proc Instit Mech Eng Part A: J Power Energy 228(3):255–271

Coxworth B (2012) Humpback whales inspire better helicopter rotor blades. New Atlas. https://newatlas.com/humpback-whales-rotor-blades/21332/

Craig GM (2002) Introduction to aerodynamics. Regenerative Press, Anderson, IN

Crisp DJ (1955) The behaviour of barnacle cyprids in relation to water movement over a surface. J Exp Biol 32:569–590

Crisp DJ, Stubbings HG (1957) The orientation of barnacles to water currents. J Anim Ecol 26:179–196

Custodio D, Henoch C, Johari H (2010) The performance of finite-span hydrofoils with humpback whale-like leading edge protuberances. 64th Annual Meeting APS Division of Fluid Dynamics, vol. 55, No. 16, 21 Nov 2010, Long Beach, CA

Darwin C (1859) On the origin of species by means of natural selection. John Murray, London

Dewar SW, Watts P, Fish FE (2013) Turbine and compressor employing tubercle leading edge rotor design. US Patent 8535008 B2, 17 Sept 2013

Dropkin A, Custodio D, Henoch CW, Johari H (2012) Computation of flow field around an airfoil with leading edge protuberances. J Aircraft 49(5):1336–1344

Edel RK, Winn HE (1978) Observations on underwater locomotion and flipper movement of the humpback whale *Megaptera novaeangliae*. Mar Biol 48:279–287

Erickson GE (1995) High angle-of-attack aerodynamics. Ann Rev Fluid Mech 27:45–88

Faber JA, Arrieta AF, Studart AR (2018) Bioinspired spring orgami. Science 359:1386–1391

Filho GB, De Costa AL, de Paula AA, De Lima GR (2018) A numerical investigation of the wavy leading edge phenomena at transonic regime. 2018 AIAA Aerospace Science Meeting

Fish FE (1998) Imaginative solutions by marine organisms for drag reduction. In: Meng JCS (ed) Proceedings of the international symposium on seawater drag reduction, Newport, Rhode Island, pp 443–450

Fish FE (2000) Limits of nature and advances of technology: What does biomimetics have to offer? In: Kato N, Suzuki Y (eds) Proceedings of the 1st international symposium on aqua bio-mechanisms/International seminar on aqua bio-mechanisms, Tokai University Pacific Center, Honolulu, Hawaii, pp 3–12

Fish FE (2002) Balancing requirements for stability and maneuverability in cetaceans. Integr Comp Biol 42(1):85–93

Fish FE (2006) Limits of nature and advances of technology in marine systems: What does biomimetics have to offer to aquatic robots? Appl Bionics Biomech 3(1):49–60

Fish FE, Battle JM (1995) Hydrodynamic design of the humpback whale flipper. J Morph 225:51–60

Fish FE, Beneski JT (2014) Evolution and bio-inspired design: natural limitations. In: Goel A, McAdams DA, Stone RB (eds) Biologically inspired design: computational methods and tools. Springer, London, pp 287–312

Fish FE, Kocak DM (2011) Biomimetics and marine technology: an introduction. Mar Tech Soc J 45(4):8–13

Fish FE, Lauder GV (2006) Passive and active flow control by swimming fishes and mammals. Ann Rev Fluid Mech 38:193–224

Fish FE, Lauder GV (2017) Control surfaces of aquatic vertebrates in relation to swimming modes. J Exp Biol 220:4351–4363

Fish FE, Rohr J (1999) Review of dolphin hydrodynamics and swimming performance. SPAWARS Systems Center Technical Report 1801, San Diego, CA

Fish FE, Howle LE, Murray MM (2008) Hydrodynamic flow control in marine mammals. Integr Comp Biol 211:1859–1867

Fish FE, Weber PW, Murray MM, Howle LE (2011a) The humpback whale's flipper: application of bio-inspired tubercle technology. Integr Comp Biol 51:203–213

Fish FE, Weber PW, Murray MM, Howle LE (2011b) Marine applications of the biomimetic humpback whale flipper. Mar Tech Soc J 45(4):198–207

Forbes P (2005) The gecko's foot. Norton, New York

Friedlaender AS, Hazen EL, Nowacek DP, Halpin PN, Ware C, Weinrich MT, Hurst T, Wiley D (2009) Diel changes in humpback whale *Megaptera novaeangliae* feeding behavior in response to sand lance *Ammodytes* spp. behavior and distribution. Mar Ecol Prog Ser 395:91–100

Ginter CC, Fish FE, Marshall CD (2010) Morphological analysis of the bumpy profile of phocid vibrissae. Mar Mamm Sci 26:733–743

Ginter CC, Böttger SA, Fish FE (2011) Morphology and microanatomy of harbor porpoise (*Phocoena phocoena*) dorsal fin tubercles. J Morph 272:27–33

Goldbogen JA, Calambokidis J, Shadwick RE, Oleson EM, McDonald MA, Hildebrand JA (2006) Kinematics of foraging dives and lunge-feeding in fin whales. J Exp Biol 209:1231–1244

Goldbogen JA, Calambokidis J, Croll DA, Harvey JT, Newton KM, Oleson EM, Schorr G, Shadwick R (2008) Foraging behavior of humback whales: kinematics and respiratory patterns suggest a high cost for a lunge. J Exp Biol 211:3712–3719

Goruney T, Rockwell D (2009) Flow past a delta wing with a sinusoidal leading edge: near-surface topology and flow structure. Exp Fluids 47:321–331

Gosline JM (1991) Efficiency and other criteria for evaluating the quality of structural biomaterials. In: Blake RW (ed) Efficiency and economy in animal physiology. Cambridge: Cambridge University Press, pp 43–64

Gould SJ (1983) Hen's teeth and horse's toes. Norton, New York

Gravish N, Lauder GV (2018) Robotics-inspired biology. J Exp Biol 221, jeb138438

Gupta A, Alsultan A, Amano RS, Kumar S, Welsh AD (2013) Design and analysis of wind turbine blades: winglet, tubercle, and slotted. In: ASME Expo 2013, turbine technical conference and exposition

Hain JHW, Carter GR, Kraus SD, Mayo CA, Winn HE (1982) Feeding behavior of the humpback whale, *Megaptera novaeangliae*, in the western North Atlantic. Fish Bull 80:259–268

Hanke W, Witte M, Miersch L, Brede M, Oeffner J, Mark M, Hanke F, Leder A, Dehnhardt G (2010) Harbor seal vibrissa morphology suppresses vortex-induced vibrations. J Exp Biol 213:2665–2672

Hansen KL, Kelso RM, Dally BB (2009) The effect of leading edge tubercle geometry on the performance of different airfoils. In: 7th World conferences experimental heat transfer, fluid mechanics thermodynamics, 28 June–03 July 2009, Krakow, Poland

Hansen KL, Kelso RM, Doolan CJ (2010) Reduction of flow induced tonal noise through leading edge tubercle modifications. In: 16th AIAA/CEAS Aeroacoustics Conference, Stockholm, Sweden, 7–9 June 2010

Hansen KL, Kelso RM, Dally BB (2011) Performance variations of leading-edge tubercles for distinct airfoil profiles. AIAA J 49:185–194

Hansen KL, Kelso RM, Doolan CJ (2012) Reduction of flow induced airfoil tonal noise using leading edge sinusoidal modifications. Acoust Aust 40:172–177

Harman J (2013) The shark's paintbrush: biomimicry and how nature is inspiring innovation. White Cloud Press, Ashland

Harris JS (1989) An airplane is not a bird. Invent Tech 5:18–22

Hazen EL, Friedlaender AS, Thompson MA, Ware CR, Weinrich MT, Halpin PN, Wiley DN (2009) Fine-scale prey aggregations and foraging ecology of humpback whales *Megaptera novaeangliae*. Mar Ecol Prog Ser 395:75–89

Hertel H (1966) Structure, form, movement. Reinhold, New York

Hoerner SF (1965) Fluid-dynamic drag. Published by author, Brick Town, NJ

Howle LE (2009) WhalePower wenvor blade. A report in the efficiency of a WhalePower Corporations. 5 meter prototype wind turbine blade. BelleQuant Engineering, PLLC

Ibrahim IH, New TH (2015) Tubercle modifications in marine propeller blades. In: 10th Pacific symposium on flow visualization and image processing. Naples, Italy, 15–18 June 2015

Ibrahim M, Alsultan A, Shen S, Amano RS (2015) Advances in horizontal axis wind turbine blade designs: introduction of slots and tubercle. J Energy Resour Tech 137:051205

Ingebrigtsen A (1929) Whales caught in the North Atlantic and other seas. Rapp P-V Reun Cons Int Explor Mer 56:1–26

Ingram J (2008) The Daily Planet book of cool ideas. Penguin, Toronto

Johari H, Henoch C, Custodio D, Levshin A (2007) Effects of leading-edge protuberances on airfoil performance. AIAA J 45:2634–2642

Jurasz CM, Jurasz VP (1979) Feeding modes of the humpback whale, *Megaptera novaeangliae*, in southeast Alaska. Sci Rep Whales Res Inst 31:69–83

Kahn A (2017) Adapt: how humans are tapping into nature's secrets to design and build a better future. St Martin's Press, New York

Kajiura SM, Forni JB, Summers AP (2003) Maneuvering in juvenile carcharhinid and sphyrnid sharks: the role of the hammerhead shark cephalofoil. Zoology 106:19–28

Katz SL, Jordan CE (1997) A case for building integrated models of aquatic locomotion that couple internal and external forces. In: Tenth international symposium on unmanned untethered submersible technolgy: proceedings of the special session on bio-engineering research related to autonomous underwater vehicles. Autonomous Undersea Systems Institute, Lee, NH, pp 135–152

Keerthi MC, Rajeshwaran MS, Kushari A, De A (2016) Effect of leading-edge tubercles on compressor cascade performance. AIAA J 54(3):1–12

Kim JW, Haeri S, Joseph P (2016) On the reduction of aerofoil–turbulence interaction noise associated with wavy leading edges. J Fluid Mech 792:526–552

Kulkani S, Chapman C, Shah H, Parn EA, Edwards DJ (2018) Designing an efficient tidal turbine blade through bio-mimicry: a systematic review. J Eng Design Tech 16(1):101–124

Larrabee EE (1980) The screw propeller. Sci Am 243:134–148

Lau ASH, Kim JW (2013) The effect of wavy leading edges on aerofoil-gust interaction noise. J Sound Vibr 332:6234–6253

Leighton T, Finfer D, Grover E, White P (2007) An acoustical hypothesis for the spiral bubble nets of humpback whales, and the implications for whale feeding. Acoustics Bull 32:17–31

Leung K (2014) Investigation of wind turbine blades with tubercles. Adv Mat Res 1051:832–839

Lewis CA (1978) A review of substratum selection in free-living and symbiotic cirripeds. In: Chia F-S, Rice ME (eds) Settlement and metamorphosis of marine invertebrate larvae. Elsevier, New York, pp 207–218

Lillienthal O (1911) Birdflight as the basis of aviation. Longmans, Green

Lohry MW, Clifton D, Martinelli L (2012) Characterization and design of tubercle leading-edge wings. ICCFD7-4302 Abstract

Lohry MW, Martinelli L, Kollassch JS (2013) Genetic algorithm optimization of periodic wing protuberances for stall mitigation. In: Proceedings of 31st AIAA applied aerodynamics conference, vol 2905, pp 1–13

Lu Y (2004) Significance and progress of bionics. J Bionics Eng 1:1–3

Luria SE, Gould SJ, Singer S (1981) A view of life. Benjamin/Cummings, Menlo Park

Madier D (2014) Aerodynamic & mechanical updates 2014, vol 2. F1-Technical Files. http://www.f1-forecast.com/pdf/F1%20Season%202014%20-%20Aerodynamic%20and%20Mechanical%20Updates%20-%20Volume%20II.pdf

Madsen CJ, Herman LM (1980) Social and ecological correlates of cetacean vision and visual appearance. In: Herman LM (ed) Cetacean behavior: mechanisms & functions. RE Krieger Publication, Malabar, FL, pp 101–147

Mai H, Dietz G, Geißler W, Richter K, Bosbach J, Richard H, de Goot K (2006) Dynamic stall control by leading edge vortex generators. American Helicopter Soc 62nd Annual Forum, Phoenix, AZ, 9–11 May 2006

McCracken GF, Safi K, Kunz TH, Dechmann DKN, Swartz SM, Wikelski M (2018) Airplane tracking documents the fastest flight speeds recorded for bats. R Soc Open Sci 3:160398

McCullough D (2015) The Wright brothers. Simon & Schuster, New York

Miklosovic DS, Murray MM, Howle LE, Fish FE (2004) Leading edge tubercles delay stall on humpback whale (*Megaptera novaeangliae*) flippers. Phys Fluids 16(5):L39–L42

Minasian SM, Balcomb KC III, Foster L (1984) The world's whales. Smithsonian Books, Washington, DC

Mueller T (2008) Biomimetics: design by nature. Nat Geo 213:68–91

Murray MM, Fish FE, Howle LE, Miklosovic DS (2005) Stall delay by leading edge tubercles on humpback whale flipper at various sweep angles. In: Proceedings of 14th international symposium unmanned untethered submersible technology. Autonomous Undersea Systems Institute, Durham, NH

Murray M, Gruber T, Fredriksson D (2010) Effect of leading edge tubercles on marine tidal turbine blades. BAPS.2010.DFD.HC.6. In: 63rd Annual Meeting APS Division Fluid Dynamics, vol 55, No. 16, Long Beach, CA, 21 Nov 2010

Nakaya K (1995) Hydrodynamic function of the head in the hammerhead sharks (Elasmobranchii: Sphyrnidae). Copeia 1995:330–336

Nishiwaki M (1972) General biology. In: Ridgway SH (ed) Mammals of the sea: biology and medicine. C C Thomas, Springfield, pp 3–204

Norberg UM (1990) Vertebrate flight: mechanics, physiology, morphology, ecology and evolution. Springer, Berlin

Oeffner J, Lauder GV (2012) Hydrodynamic function of shark skin and two biomimetic applications. J Exp Biol 215:785–796

Owen JC, Szewezyk AA, Bearman PW (2000) Suppression of karman vortex shedding. Gallery of fluid motion. Phys Fluids 12(9):1–13

Ozen CA, Rockwell D (2010) Control of vortical structures on a flapping wing via a sinusoidal leading-edge. Phys Fluids 22:021701

Paterson EG, Wilson RV, Stern F (2003) General-purpose parallel unsteady RANS CFD code for ship hydrodynamics. IIHR Hydroscience and Engineering Report 531. The University of Iowa, Iowa City, IA

Pedro HTC, Kobayashi MH (2008) Numerical study of stall delay on humpback whale flippers. In: 46th AIAA Aerospace Meeting Exhibit, 7–10 Jan 2008, Reno, Nevada

Petrin CE, Elbing BR, Martin T, Caire W, Thies ML (2018) Modification of drag on the ear of Brazilian free-tailed bats (*Tadarida brasiliensis*) via leading-edge tubercles. In: 2018 Fluid dynamics conference, AIAA Aviation Forum (AIAA 2018-2917)

Pivorunas A (1979) The feeding mechanisms of baleen whales. Am Sci 67:432–440

Polacsek C, Reboul G, Clair V, Le Garrec T, Deniau H (2011) Turbulence-airfoil interaction noise reduction using wavy leading edge: an experimental and numerical study. Proc Inter-Noise Congress 2011(2):4464–4474

Potvin J, Goldbogen JA, Shadwick RE (2009) Passive versus active engulfment: verdict from trajectory simulations of lunge-feeding fin whales *Balaenoptera physalus*. J R Soc Interface 6:1005–1025

Potvin J, Goldbogen JA, Shadwick RE (2012) Metabolic expenditures of lunge-feeding rorquals across scale: implications for the evolution of filter-feeding and the limits to maximum body size. PLoS ONE 7:e44854

Quinn S, Gaughran W (2010) Bionics—an inspiration for intelligent manufacturing and engineering. Robot Comput Integr Manuf 26(6):616–621

Ralston E, Swain G (2009) Bioinspiration-the solution for biofouling control. Bioinsp Biomim 4(1):015007

Reidenberg JS, Laitman JT (2007) Blowing bubbles: An aquatic adaptation that risks protection of the respiratory tract in humpback whales (*Megaptera novaeangliae*). Anat Rec 260:569–580

Ridgway SH, Harrison R (1985) Handbook of marine mammals, vol. 3: the sirenians and baleen whales. Academic Press, London

Rostamzadeh N, Hansen KL, Kelso RM, Dally BB (2014) The formation mechanism and impact of streamwise vortices on NACA 0021 airfoil's performance with undulating leading edge modification. Phys Fluids 26(11):107101

Saadat M, Haj-Hariri H, Fish F (2010) Explanation of the effects of leading-edge tubercles on the aerodynamics of airfoils and finite wings. 63rd Annual Meeting APS Division Fluid Dynamics, vol 55, No. 16, 21 Nov 2010, Long Beach, CA

Sandin RCL (2014) Aircraft tail surface with a leading edge section of undulating shape. US patent 8789793 B2, 29 July 2014

Segre PS, Cade DE, Fish FE, Allen AN, Calambokidis J, Friedlander AS, Goldbogen JA (2016) Hydrodynamic properties of fin whale flippers predict rolling performance. J Exp Biol 219:3315–3320

Serson D, Meneghini JR, Sherwin SJ (2017) Direct numerical simulations of the flow around wings with spanwise waviness at a very low Reynolds number. Comput Fluids 146:117–124

Shadwick RE, Goldbogen JA, Potvin J, Pyenson ND, Vogl AW (2013) Novel muscle and connective tissue design enables high extensibility and controls engulfment volume in lunge-feeding rorqual whales. J Exp Biol 216:2691–2701

Sharpe FA, Dill LM (1997) The behavior of Pacific herring schools in response to artificial humpback whale bubbles. Can J Zool 75:725–730

Shevell RS (1986) Aerodynamic anomalies: can CFD prevent or correct them? J Aircraft 23:641–649

Shi W, Rosli R, Atlar M, Norman R, Wang D, Yang W (2016) Hydrodynamic performance evaluation of a tidal turbine with leading-edge tubercles. Ocean Eng 117:246–253

Shi W, Atlar M, Norman R (2017) Detailed flow measurement of the field around tidal turbines with and without biomimetic leading-edge tubercles. Renew Energ 111:688–707

Shi W, Atlar M, Norman R (2018) Learning from humpback whales for improving the energy capturing performance of tidal turbine blades. In: Ölcer AI, Kitada M, Dalaklis D, Ballini F (eds) Trends and challenges in maritime energy management. Springer, Cham, pp 479–497

Stanway MJ (2008) Hydrodynamic effects of leading-edge tubercles on control surfaces and in flapping foil propulsion. MS thesis, Massachusetts Institute of Technology, Cambridge, MA

Taubes G (2000) Biologists and engineers create a new generation of robots that imitate life. Science 288:80–83

Tomilin AG (1957) Mammals of the U.S.S.R. and adjacent countries. Vol. IX: Cetacea. Nauk S.S.S.R Moscow (English Translation, 1967, Israel Program for Scientific Translations, Jerusalem)

True FW (1983) The whalebone whales of the Western North Atlantic. Smithsonian Institution Press, Washington, DC

Turner JM, Kim JW (2017) Aeroacoustic source mechanisms of a wavy leading edge undergoing vortical disturbances. J Fluid Mech 811:562–611

van Nierop EA, Alben S, Brenner MP (2008) How bumps on whale flippers delay stall: an aerodynamic model. Phys Rev Lett 100:054502-1–054502-4

Vaughan TA (1986) Mammalogy. Sauders, Philadelphia

Velcro SA (1955) Improvements in or relating to a method and a device for producing velvet type fabric. Switzerland Patent 721338

Vincent J (1990) Structural biomaterials. Princeton University Press, Princeton, NJ

Vogel S (1981) Life in moving fluids. Willard Grant Press, Boston

Vogel S (1988) Life's devices. Princeton University Press, Princeton

Vogel S (1994) Life in moving fluids. Princeton University Press, Princeton

Vogel S (1998) Cat's paws and catapults. WW Norton, New York

von Mises R (1945) Theory of flight. Dover, New York

Walsh MJ (1990) Riblets. Prog Astro Aero 123:203–261

Wang J, Zhang C, Wu Z, Wharton J, Ren L (2017) Numerical study on reduction of aerodynamic noise around an airfoil with biomimetic structures. J Sound Vib 394:46–58

Watts P, Fish FE (2001) The influence of passive, leading edge tubercles on wing performance. In: Proceedings of twelfth international symposium unmanned untethered submersible technology. Autonomous Undersea Systems Institute, Durham, NH

Watts P, Fish FE (2002) Scalloped wing leading edge. US Patent 6,431,498, 13 Aug 2002

Webb PW (1997) Designs for stability and maneuverability in aquatic vertebrates: What can we learn? In: Proceedings of tenth international symposium unmanned untethered submersible technology: proceedings of the special session on bio-engineering research related to autonomous underwater vehicles. Autonomous Undersea Systems Institute, Lee, NH, pp 86–103

Weber PW, Howle LE, Murray MM, Fish FE (2009) Lift and drag performance of odontocete cetacean flippers. J Exp Biol 212:2149–2158

Weber PW, Howle LE, Murray MM (2010) Lift, drag and cavitation onset on rudders with leading edge tubercles. Mar Tech 47:27–36

Weber PW, Howle LE, Murray MM, Miklosovic DS (2011) Computational evaluation of the performance of lifting surfaces with leading edge protuberances. J Aircraft 48:591–600

Weber PW, Howle LE, Murray MM, Reidenberg JS, Fish FE (2014) Hydrodynamic performance of the flippers of large-bodied cetaceans in relation to locomotor ecology. Mar Mamm Sci 30:413–432

Wegener PP (1991) What makes airplanes fly. Springer, New York

Weihs D (1981) Effects of swimming path curvature on the energetics of fish swimming. Fish Bull 79:171–176

Weihs D (1993) Stability of aquatic animal locomotion. Cont Math 141:443–461

Whitehouse G (2011) Turbocled tubercles. SeaHorse 2011:28–29

Wiley D, Ware C, Bocconcelli A, Cholewiak D, Friedlaender A, Thompson M, Weinrich M (2011) Underwater components of humpback whale bubble-net feeding behaviour. Behaviour 148:575–602

Wind Energy Institute of Canada (2008) WhalePower tubercle blade power performance testreport. Wind Energy Institute of Canada, North Cape

Winn HE, Reichley NE (1985) Humpback whale *Megaptera novaeangliae* (Borowski, 1781). In: Ridgway SH, Harrison R (eds) Handbook of marine mammals, vol 3. The sirenians and baleen whales. Academic Press, London, pp 241–273

Winn LK, Winn HE (1985) Wings in the sea: the humpback whale. University Press of New England, Hanover

Woodward BL, Winn JP, Fish FE (2006) Morphological specializations of baleen whales associated with hydrodynamic performance and ecological niche. J Morph 267:1284–1294

Wu JZ, Vakili AD, Wu JM (1991) Review of the physics of enhancing vortex lift by unsteady excitation. Prog Aerosp Sci 28:73–131

Xingwei Z, Chaoying Z, Tao Z, Wenying J (2013) Numerical study on effect of leading-edge tubercles. Aircraft Eng Aerospace Tech 85:247–257

Zangerl R, Case GR (1973) Iniopterygia, a new order of chondrichthyan fishes from the Pennsylvanian of North America. Fieldiana Geol Mem 6:1–67

Zhang RK, Wu JZ, Chen SY (2012) A new active control strategy for wind-turbine blades under off-design conditions. Int J Mod Phys Conf Ser 19:283–292

Zheng T, Qiang X, Teng J, Feng J (2018) Investigation of leading edge tubercles with different wavelengths in an annular compressor cascade. Int J Turbo Jet Eng 2018-02-09

Zibkoff M (2009) Spoked bicycle wheel. US Patent Application US2009/0236902, 19 Mar 2009

Tubercled Wing Flow Physics and Performance

N. Rostamzadeh, K. Hansen and R. Kelso

1 Introduction

This chapter provides a review of the current understanding of both the potential performance effects conferred by the application of tubercles as well as our understanding of the mechanisms by which these effects are produced. While it is clear that tubercles can both beneficially and detrimentally affect performance, depending on the flow conditions, the mechanism by which this occurs and the influence of Reynolds number remain active topics of debate. The following discussion describes nine different flow mechanisms that have been proposed to explain the observed performance effects, as well as our state of knowledge of Reynolds number effects. The influence of tubercle geometric parameters such as the amplitude-to-wavelength ratio (A/λ) and amplitude-to-chord ratio (A/c) are shown to be important. Finally, the effects of tubercles on wings of finite aspect ratio (AR) and sweep, both of which influence the large-scale three-dimensionality of the flow, are also discussed.

Ever since the publication of the seminal paper "Hydrodynamic Design of the Humpback Whale Flipper," by Fish and Battle (1995), numerous mechanisms have been proposed to cast light on the influence of leading-edge (LE) rounded tubercles on the flow field and the performance of lifting-bodies that incorporate similar features. In their work, Fish and Battle (1995) suggested that LE protuberances, as observed on humpback flippers, render hydrodynamic advantages by maintaining lift at high attack angles, thus contributing to the agility of the humpback whale. Subsequently, Miklosovic et al. (2004), through a series of wind tunnel tests, confirmed that a tubercled flipper prototype outperformed a flipper model without tubercles by extending the stall angle by 40% as well as increasing the maximum lift coefficient

N. Rostamzadeh · R. Kelso (✉)
School of Mechanical Engineering, University of Adelaide, Adelaide 5005, Australia
e-mail: richard.kelso@adelaide.edu.au

K. Hansen
College of Science and Engineering, Flinders University, Adelaide 5042, Australia

© Springer Nature Switzerland AG 2020
D. T. H. New and B. F. Ng (eds.), *Flow Control Through Bio-inspired Leading-Edge Tubercles*, https://doi.org/10.1007/978-3-030-23792-9_2

by 6%. The Reynolds number (*Re*, based on chord length) pertaining to these tests was maintained at 505,000 to 520,000 which corresponds to the normal flow regime of the humpback whale. This study corroborated the theory that under certain flow conditions LE tubercles can yield remarkable performance benefits.

2 Flow Physics

To date, various effects of tubercle-like LE modifications on the flow field have been explored through experimental, numerical and theoretical approaches. Nonetheless, competing proposed mechanisms have yet to reach consensus on the underlying flow physics activated by the presence of these undulating protuberances. What follows is a review of the more compelling hypotheses addressing the roles played by tubercles.

2.1 Vortex Generators

Miklosovic et al. (2004) postulated that tubercles might act in a similar fashion to conventional vortex generators utilized on aircraft wings. It was suggested that tubercles produce vortices that 'energize' the boundary layer developed over the lifting-body and inhibit chordwise flow separation, leading to delayed stall. It must be noted that the height of a typical vortex generator is in the same order of magnitude as the boundary layer thickness (Lin 2002). The role of tubercles as vortex generators was subsequently challenged by van Nierop et al. (2008), who argued that, as opposed to the dimensions of a vortex generator, the amplitude and wavelength associated with LE tubercles are much larger than the boundary layer thickness. Hansen et al. (2011), however, showed that the value of a newly-defined parameter called the effective height ratio (equal to tubercle amplitude multiplied by the sine of the angle of attack (AoA) divided by the boundary layer thickness) is on par with the height ratio in a situation where a vortex generator is used. Zhang et al. (2014) demonstrated that, similar to micro vortex generators, the tubercle height ratio is 0.1–0.5. Through flow visualization studies, Hansen et al. (2011) observed the presence of counter-rotating streamwise vortices similar to those produced by conventional vortex generators. The streamwise vortices generated between tubercles increased in size and decreased in intensity as they travelled over the modified lifting body. Furthermore, Hansen et al. (2011) proposed that the tubercle amplitude-to-wavelength ratio is an important geometric parameter affecting the strengths of the streamwise vortices. Consistent with this notion, a numerical study by Rostamzadeh et al. (2014) suggested that airfoils with higher amplitude-to-wavelength ratios might generate stronger counter-rotating vortices.

In order to investigate the previously proposed hypothesis that tubercles act as vortex generators, Stein and Murray (2005) carried out wind tunnel tests on airfoils based on the NACA0020 profile at $Re = 250,000$. They compared the performance of a tubercle-modified airfoil with an unmodified one, as well as an airfoil with vane vortex generators. It was found that maximum lift coefficient of the modified airfoil was lower than that of the baseline. Furthermore, the lift and drag curves associated with the modified airfoil deviated from those of the airfoil fitted with vortex generators. Thus, it was concluded that tubercles do not merely serve as conventional vortex generators. In spite of this conclusion, Bolzon et al. (2015) noted that while the geometrical dimensions and spacing of the vortex generators were optimized in the aforementioned study, those of the tubercles were not. This might explain the observed discrepancy between the aerodynamic behavior of the tubercled airfoil and the one fitted with vortex generators (Bolzon et al. 2015).

The vortex-generating capabilities of tubercles were further highlighted by a study carried out on a swept wing operating at pre-stall AoAs. Bolzon et al. (2016) demonstrated that, at $Re = 225,000$, the effects of sweep are reflected in the strengths of the counter-rotating streamwise vortices. Accordingly, in a pair of tubercle-generated vortices, one vortex may become as much as four times as strong as its counter-rotating partner. It was also reported that the strength of the wingtip vortex is reduced when the AoA exceeds 6°, conditions under which flow separation occurs near the wingtip.

2.2 Rounded Delta Wings

Custodio (2007) posited that each tubercle could be conceived of as a small-scale delta wing, producing counter-rotating streamwise vortices on either side of the tubercle peak (Fig. 1). It was also conjectured that a tubercled wing may produce vortex lift in an analogous manner to a delta wing. Bolzon et al. (2015), however, challenged this

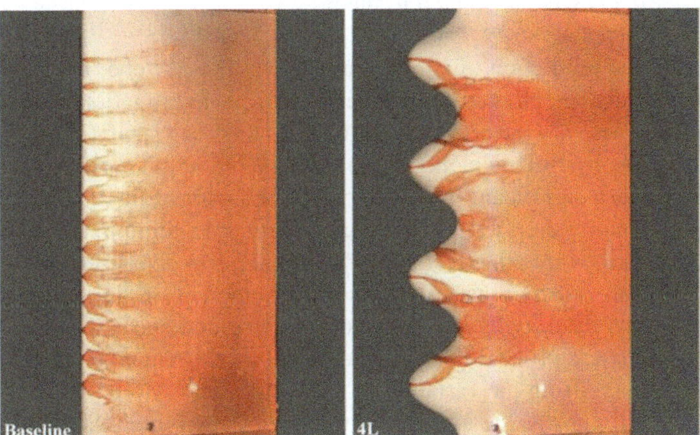

Fig. 1 Tubercled wings produce pairs of streamwise vortices in a bi-periodic manner as demonstrated in the flow visualization work by Custodio (2007)

idea by pointing out that while the counter-rotating vortices of a delta wing produce downwash over the surface of the wing, the upwash does not occur immediately above the wing. Therefore, whereas the high-lift generation capabilities of a delta wing are brought about primarily by the downwash effect, this is not the case with the flow over a tubercled wing whereby the streamwise vortices produce both upwash and downwash immediately above the wing. This implies that the lift produced by a tubercled wing is affected both by flow separation induced by the upwash as well as the flow attachment aided by the downwash phenomena. A second difference was attributed to the strengths of the vortices generated by tubercles which are expected to be lower in comparison with those of a delta wing owing to the smaller geometric dimensions of tubercles.

2.3 *Variation of the Effective Angle-of-Attack*

van Nierop et al. (2008) argued that as the airfoil cross-sectional thickness associated with a tubercle peak is approximately the same as that of a trough, the difference in chord lengths gives rise to a stronger adverse pressure gradient behind the trough compared with that behind the peak. In addition, it was demonstrated that the effective AoA, α, varies along the span of a tubercled wing such that the local AoA over a tubercle peak is smaller than that of a tubercle trough. This difference in the effective AoA was attributed to the downwash effect induced over and behind the peak and conversely the upwash effect over the trough. van Nierop et al. (2008) proposed that the downwash effect over the peak, along with a weaker adverse pressure gradient in that region, result in more attached flow behind tubercle peaks. A Particle-Image Velocimetry (PIV) study by Hansen et al. (2016) confirmed the downwash and upwash effect behind tubercled peaks and troughs, respectively, as illustrated in Fig. 2. These effects were attributed to the presence of counter-rotating streamwise vortices, as predicted.

In line with the prediction of van Nierop et al. (2008), Weber et al. (2011) provided further evidence obtained from their numerical investigation to reinforce the theory that tubercles vary the effective AoA across the wing span. Furthermore, flow visualization work by Bolzon et al. (2017) illustrated that the presence of a varying effective AoA has a significant influence on the behavior of the laminar separation bubbles (LSBs), formed on a swept tubercled wing at small pre-stall AoAs. It was elucidated that the downwash effect behind tubercle peaks stabilized the flow, delaying the process of transition as well as the re-attachment of the LSB. In contrast, upwash over tubercle troughs destabilized the separated shear layer and accelerated the re-attachment process in this region. Also consistent with this mechanism, induced drag arising from the variation in the effective AoA was found to be lower behind tubercle troughs as the apparent tilting of the lift vector was reduced due to upwash in the trough region (Bolzon et al. 2017).

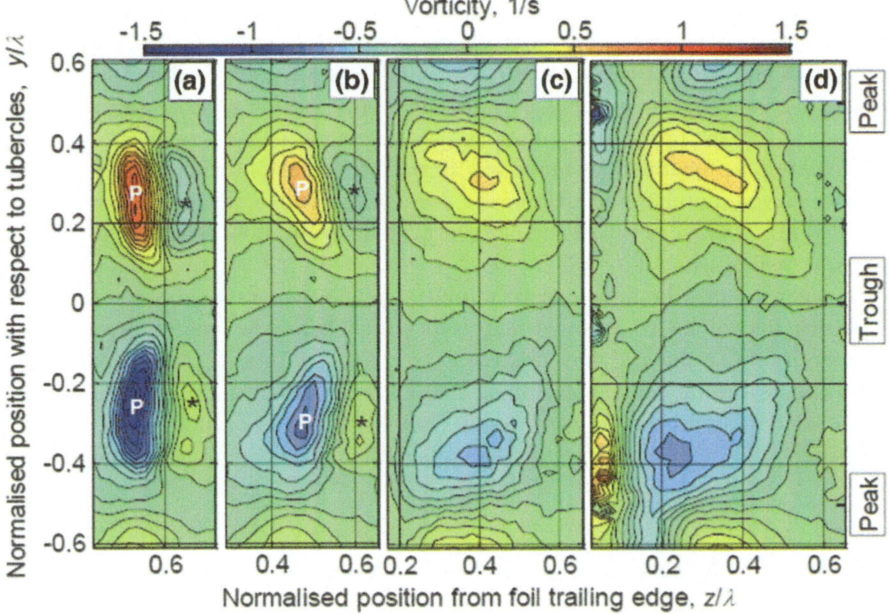

Fig. 2 Counter-rotating pairs of vortices create upwash above the trough and downwash above the peak spanwise locations as evidenced by the vorticity field in the PIV-based work by Hansen et al. (2016). Red and blue indicate regions where vorticity assumes a counter-clockwise and clockwise direction, respectively

2.4 Spanwise Progression of Stall

A number of researchers (Fish and Battle 1995; Stein and Murray 2005; Miklosovic et al. 2007; Pedro and Kobayashi 2008) have suggested that tubercles may inhibit the progression of spanwise flow that often leads to tip stall on finite-wings by compartmentalizing the flow field. In their numerical work, Watts and Fish (2001) argued that counter-rotating vortices provide barriers to spanwise flow, in a similar fashion to wing fences.

Further insight into the flow physics came from Miklosovic et al. (2007) who carried out wind tunnel tests on flipper models at $Re = 534,000$ to $631,000$ and on full-span unswept airfoils at $Re = 274,000$ to $277,000$. This investigation revealed that the flipper model with tubercles produced a higher maximum lift coefficient, yet the performance of unswept full-span airfoils with tubercles was degraded in the pre-stall regime. Hence, these researchers concluded that tubercles on flipper models render aerodynamic benefits by inhibiting the spanwise progression of stall. It must be noted that tubercles on full-span foils still outperformed the unmodified foil in the post-stall regime. This observation was supported by the experimental work by Johari et al. (2007), Hansen et al. (2011) and Custodio et al. (2015).

To isolate the effect of three-dimensional flow features on the performance of unswept wings, Hansen et al. (2010) conducted wind tunnel tests on finite-span wings with and without tubercles at $Re = 120,000$. It was observed that finite-span wings exhibited the same trends as full-span airfoils tested at the same Reynolds number, i.e., in the pre-stall regime tubercles degraded the performance, yet in the post-stall regime the tubercles improved the performance. Observing these common traits displayed by tubercled airfoils and finite wings led to the conclusion that the effect of protuberances on the tip vortex, present in the flow over unswept finite wings, is insignificant at $Re = 120,000$.

By means of oil-film flow visualization, Bolzon (2017) provided compelling evidence that, in contrast to an unmodified swept wing where flow separation progresses from the trailing-edge (TE) towards the LE, on a swept tubercled wing flow separation occurs behind troughs asymmetrically. It was clearly shown that separation behind each tubercle was isolated from its neighbouring separation zones. In other words, tubercles appeared to compartmentalize the flow as previously suggested.

2.5 Varying Spanwise Circulation

Rostamzadeh et al. (2013) adopted a theoretical approach, which led to the conclusion that by deforming the LE of a wing in an undulating manner, the chordwise circulation generated at each spanwise location is altered. In their subsequent study (Rostamzadeh et al. 2014), it was argued that since chordwise circulation is the surface integral of spanwise vorticity, according to Stokes' theorem, spanwise vorticity must vary in a similar manner to circulation in the spanwise direction. One effect of an undulating LE is, therefore, to curve the apparent spanwise vortex sheets. Further evidence was corroborated by Cai et al. (2015) and Bolzon (2017) to strengthen the hypothesis that tubercles vary the wing's circulation in an undulating fashion along the span.

2.6 Enhanced Momentum and Energy Transfer

Skillen et al. (2014) adopted numerical methods to investigate the effect of tubercles on post-stall lift production in the transitional flow regime. It was reported that a secondary flow mechanism was responsible for the transfer of momentum to regions behind tubercle peaks. In another numerical investigation, Rostamzadeh et al. (2014) showed that, even at pre-stall attack angles, a pair of counter-rotating streamwise vortices was formed due to flow skewness near the LE. This flow feature was identified as Prandtl's secondary flow of the first kind. In the post-stall zone, the vortices were suggested to enhance momentum transfer through which fluid particles are transported from separated flow regions to neighbouring locations. For certain configurations at post-stall AoA, the action of the transport of momentum was carried out in a bi-periodic manner whereby the flow separation pattern repeated itself at every second trough. This sub-harmonic behavior had previously been detected in the flow visualisation work by Custodio (2007) and later by de Paula (2016). Dropkin et al. (2012) also identified bi-periodicity by observing the convergence and divergence of pathlines at adjacent troughs.

Similarly, Malipeddi et al. (2012) hypothesized that the interactions between counter-rotating vortices forming on protuberances gave rise to this sophisticated flow pattern. Camara and Sousa (2013) later found that bi-periodicity only appears for certain LE configurations in the post-stall regime. In a comprehensive numerical study, Zhao et al. (2017) reported that immediately prior to stall, the counter-rotating vortices moved away from each other in alternate troughs and lowered the LE suction in these regions. In contrast, in the post-stall regime the induced velocity associated with vortices aided the process of flow attachment over tubercle peaks.

Perez-Torro and Kim (2017) carried out numerical modelling of the flow past deeply-stalled tubercled and baseline airfoils operating at $Re = 120,000$. These researchers reported the effect of the number of LE tubercle wavelengths (i.e. the spanwise domain size) on the emerging flow patterns (Fig. 3). It was found that on the suction side of the tubercled foil, a cluster of Laminar Separation Bubbles (LSBs) was formed which grew in size with an increase in the number of tubercle wavelengths from two to four. Whereas, downstream of at least one trough the flow was completely separated, the flow remained attached over a portion of the span behind the collocated LSBs. Furthermore, pairs of streamwise vortices were suggested to serve as a buffer layer between the group of LSBs and the adjacent separated shear layer. The buffer layer prevented the penetration of the separated flow into the LSB group. The longevity of the streamwise vortices was maintained by extraction of turbulent kinetic energy from the separated shear layer.

Fig. 3 The separated and attached flow pattern on the suction side of tubercled LE as studied by Perez-Torro and Kim (2017)

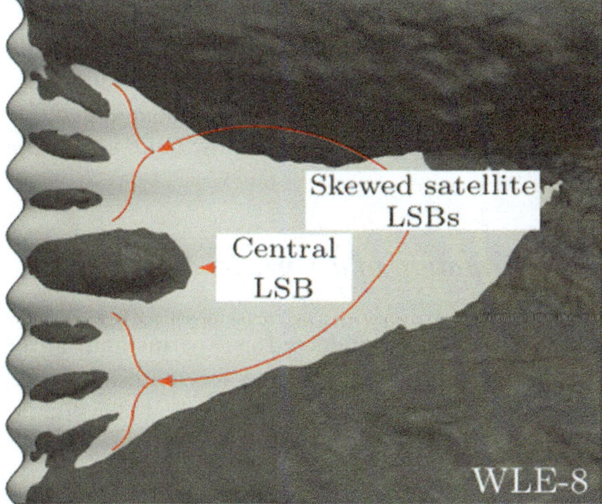

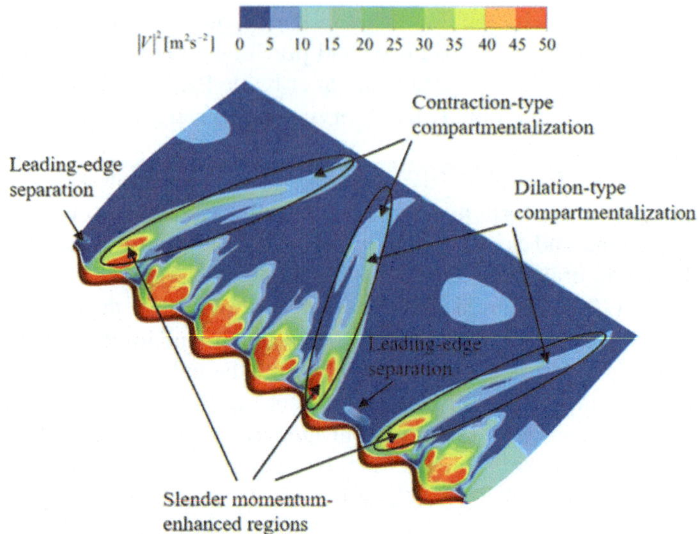

Fig. 4 Tubercles enhance momentum transfer and compartmentalize the flow as displayed in the numerical work by Cai et al. (2017)

Cai et al. (2017) undertook experiments and numerical modelling to explore the flow field around tubercled airfoils based on the NACA 63_4-021 profile at $Re = 180{,}000$. Flow visualization studies and numerical simulations revealed periodic flow patterns at small AoAs. In contrast, at higher AoAs aperiodicity was detected. It was reported that the distance between two locally-stalled trough regions was in the range of 4–7 tubercle wavelengths. To account for the aperiodic flow patterns, Cai et al. (2017) suggested the compartmentalizing effect of slender momentum-enhanced regions in close proximity to tubercle peaks. Drawing on numerical simulations, it was shown that there exist regions near the tubercled LE where momentum is enhanced through the action of streamwise vortices. Past a critical AoA, these regions assume slender shapes oriented in a skewed fashion, thus rendering a compartmentalizing effect to the flow field, as illustrated in Fig. 4.

2.7 Formation of a Vorticity Canopy

Hansen et al. (2016) conducted an experimental and a numerical study on a tubercled and an unmodified airfoil at a low Reynolds number of 2230. Drawing on the observations from both approaches, these researchers demonstrated that the tubercled LE produced strong spanwise pressure gradients which gave rise to the generation of streamwise vorticity. Streamwise vorticity was then immediately released into the flow field in a highly three-dimensional manner which subsequently re-organized itself in the shape of an apparent vorticity canopy, as illustrated in Fig. 5. At the

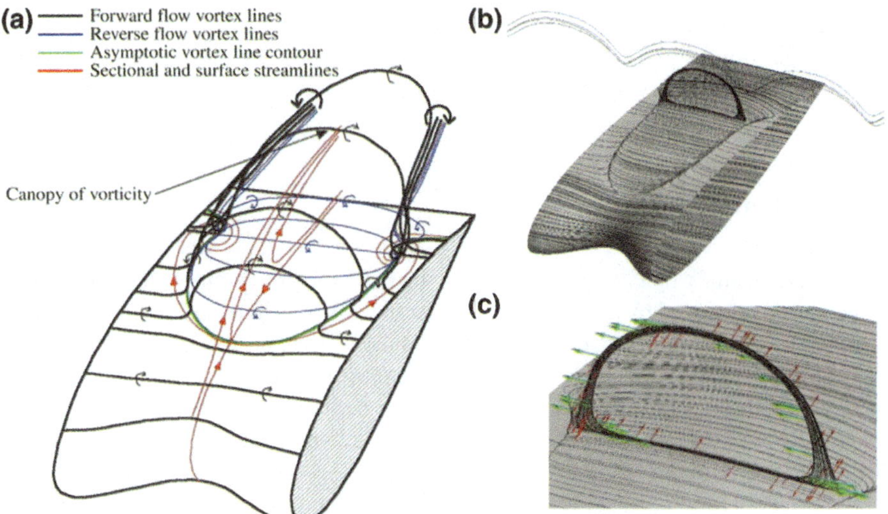

Fig. 5 Three-dimensional vortex line interpretation and corresponding computed vortex lines based on CFD simulations. Green and red arrows indicate the spanwise and streamwise orientation of vorticity respectively at the given location. $Re = 2230$. After Hansen et al. (2016)

edges of the vorticity canopy, boundary-layer vorticity entrainment increased chordwise circulation from the leading to the TE. On the sides of the vorticity canopy a pair of streamwise vortices was formed whose development was reminiscent of an owl-face flow feature.

2.8 Weakening of the Suction Peak

Serson and Meneghini (2015) conducted a computational study employing the Direct Numerical Simulation approach to investigate the flow mechanism triggered by LE tubercles on NACA0012 based airfoils at $Re = 1000$. The numerically-obtained results suggested that tubercles impede the ability of the lifting body to accelerate the flow in the chordwise direction and result in the reduction of the maximum value of the suction pressure over the tubercle peak. Serson and Meneghini (2015) proposed that as the flow approaches the LE on the suction side, the effects of low suction pressure are felt initially over the peak spanwise location (due to the extended chord in that region). This sets up a spanwise pressure gradient which accelerates the flow towards the peak spanwise location. It was argued that the spanwise movement of fluid reduces the suction peak which leads to a weaker adverse streamwise pressure gradient along the peak compared to that of the trough. Hence, flow separation occurs primarily behind tubercle troughs. In a second study, Serson et al. (2017) explored the mechanism activated by tubercles in the flow regime where $Re = 10,000$ and

50,000, respectively. Adopting the same baseline airfoil (NACA0012), the numerical simulations showed that the weakening of the suction peak over the chord maxima regions was observed even at $Re = 50,000$.

2.9 Flow Instability

In a numerical study by Favier et al. (2011), the formation of streamwise vortices was suggested to be caused by a Kelvin-Helmholtz instability when the Reynolds number was maintained at 800. Induced by this instability, tubes of vorticity initially were illustrated to orient themselves orthogonal to the tubercled wing's surface before turning into the streamwise direction and being convected by the flow. Previously Stanway (2008) had detected the presence of surface normal vorticity in a series of PIV-based experiments.

The effect of tubercles on the stability of the shear layer was further illuminated by the aforementioned numerical work of Serson et al. (2017). The simulations showed that tubercles enhance the process of flow transition. At $Re = 10,000$ where the flow was observed to be entirely laminar over the unmodified NACA0012 airfoil, the early onset of transition brought about by tubercles increased the lift coefficient of the tubercled airfoil. In contrast, at $Re = 50,000$ the occurrence of early transition over the modified airfoil led to a shortening of the separation bubble that did not yield higher values of the lift coefficient.

3 The Effects of Reynolds Number

The impact of flow regime on the mechanism activated by wing LE modifications with tubercles or tubercle-like features has been of keen interest. A pioneering study by Stanway (2008) provided evidence to demonstrate that varying the Reynolds number can influence the performance of a tubercled flipper model. In this experimental investigation carried out over a range of low $Re = 44,000$ to 120,000, the tubercled flipper produced consistently lower maximum lift coefficients than the one with the smooth LE, except for the case where $Re = 120,000$. It was posited that in addition to the type of stall (LE versus TE stall) Reynolds number effects alter the performance of tubercled wings.

In an experimental study, Hansen et al. (2010) observed that tubercled finite-span wings and full-span airfoils exhibited the same trends with regard to aerodynamic loading at the same $Re = 120,000$. Similar to airfoils, finite-span wings produced lower maximum lift coefficients compared to their unmodified counterparts. Conversely, in the post-stall zone, the lift coefficients of the modified wings and airfoils were enhanced. Typical results are presented in Fig. 6. Based on these findings, it was inferred that flow regime is likely to play a more influential role in the performance

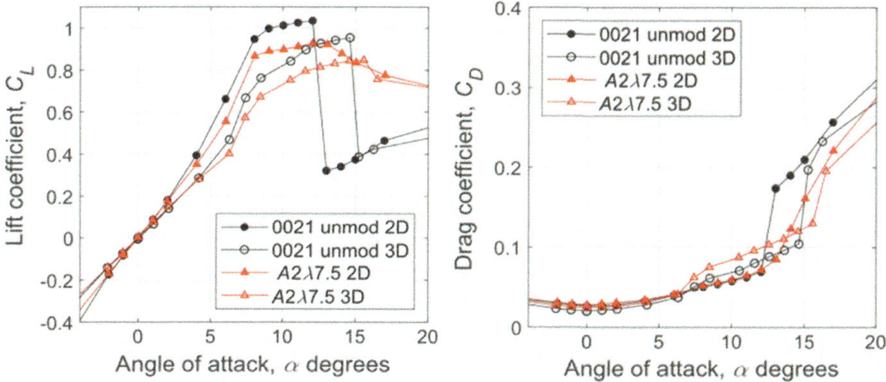

Fig. 6 Lift coefficient (left) and drag coefficient (right) for a NACA 0021 airfoil at $Re = 120,000$ as reported by Hansen et al. (2010). Comparison is made between unmodified and modified full-span (2D) and finite-span (3D) airfoils, where $A2\lambda7.5$ ($A/c = 0.029$, $\lambda/c = 0.11$) denote the amplitude, A, and wavelength, λ, in mm of the sinusoidal tubercles

of tubercled lifting bodies than three-dimensional flow features such as the tip vortex present in the flow over finite wings.

An additional observation from Fig. 6 is the variation in the slope of the lift curve, which is greatest for the unmodified 2D case and exceeds the theoretical maximum value of 2π (rad^{-1}). Hansen et al. (2014) explains this in terms of the formation of a LSB on the suction surface of the airfoil, which alters the effective camber of the airfoil. This phenomenon commonly occurs in transitional flow and is hence highly Reynolds number dependent. In Fig. 6, it is seen that the addition of tubercles decreases but does not completely eliminate the observed camber effect.

A numerical study was carried out by Dropkin et al. (2012) in which the aerodynamic characteristics of unmodified and tubercled NACA63$_4$-021 airfoils were compared over a range of $Re = 180,000$ to $3,000,000$. These researchers reported that with the increase in Reynolds number, the maximum lift-to-drag ratios achieved by both unmodified and tubercled airfoils increase. Nonetheless, whereas the maximum lift coefficient achieved by the unmodified airfoil increases, the lift coefficients of the modified airfoils remain nearly constant as the Reynolds number climbs, as shown in Fig. 7. In addition, the overall effects of Reynolds number on lift and drag characteristics of the unmodified airfoils are more pronounced than on the tubercled airfoils. It must be emphasized that these predictions based on numerically-obtained findings are yet to be verified by experimental data.

Weber et al. (2010) conducted a series of water tunnel tests on rudders with LE tubercles at $Re = 200,000$ to $800,000$. It was revealed that when operating at lower Reynolds numbers, the modified rudders' performance was degraded for AoAs between $15°$ and $20°$. Tubercles were reported to accelerate the onset of cavitation and influence the location of cavitation. With the increase in Reynolds number, however, the disparate performance of smooth and tubercled rudders diminished. The study

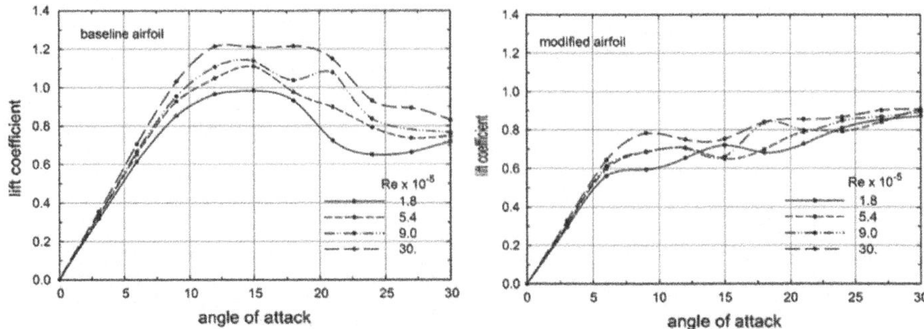

Fig. 7 The lift coefficient versus the attack angle associated with the unmodified and tubercled foils over a range of Reynolds numbers, as reported by Dropkin et al. (2012)

by Weber et al. (2010) suggested that beyond a critical Reynolds number, the effect of tubercles on performance may be insignificant.

Custodio et al. (2015) assessed the hydrodynamic performance of several finite-span wings (based on the NACA63$_4$-021 profile) against their unmodified equivalents. The tested lifting bodies consisted of seven wings with rectangular planforms, two swept wings and two models resembling the humpback whale flipper. The experiments were conducted up to $Re = 450,000$. These researchers found that, apart from the flipper models positioned at AoAs between 17° and 22°, all the other wings yielded comparable or lower lift-to-drag ratios than those achieved by their unmodified counterparts. In addition, the respective loading behavior of the rectangular-planform and swept wings was independent of the Reynolds number above $Re = 360,000$. For the flipper models, the Reynolds number beyond which the loading characteristics were not considerably affected was reported to be 180,000. These findings echo the conclusion by Weber et al. (2010) that there exists a range of the Reynolds number where the influence of the Reynolds number on loading characteristics is of significance.

The influence of Reynolds number was also highlighted in the numerical study by Rostamzadeh et al. (2017) whereby the flow mechanisms associated with a tubercled and an unmodified airfoil (NACA0021) were assessed at $Re = 120,000$ and 1,500,000. This numerical investigation, suggested that the post-stall superior performance of tubercled airfoils in the transitional flow regime, as reported in the literature (Hansen et al. 2011; Johari et al. 2007), may be absent for the near-turbulent conditions. It was argued that, in contrast to flow conditions at lower Reynolds numbers, in the near-turbulent zone the LSB forming on the unmodified foil becomes more resistant to bursting and the possibility of a sudden loss of lift is precluded. Thus, the airfoil with the smooth LE is likely to produce more lift in the post-stall zone than the tubercled airfoil. It must be noted that further experimental evidence is required to substantiate these findings.

4 Unswept Full-Span Wing Performance

The presence of tubercles on the LE of an airfoil has shown to improve aerodynamic performance (Watts and Fish 2001; Miklosovic et al. 2004; Pedro and Kobayashi 2008) through separation delay and increased maximum stall angle (Miklosovic et al. 2004). However, performance degradation has also been observed in many studies (Stein and Murray 2005; Johari et al. 2007; Hansen et al. 2011), at AoA prior to stall. Post-stall, LE tubercles have been shown to increase lift generation (Johari et al. 2007; Hansen et al. 2011; Yoon et al. 2011; Skillen et al. 2014) and in some cases, reduce drag as well (Miklosovic et al. 2004; Hansen et al. 2011; Rostamzadeh et al. 2013). It is therefore useful to compare studies that investigated the performance of the simplest airfoil configuration, which is a full-span, unswept wing.

Stein and Murray (2005) demonstrated that a NACA0020 airfoil with tubercles produced reduced lift and increased drag compared with the unmodified airfoil. Experiments were carried out for $0° \leq \alpha \leq 12°$ at $Re = 250{,}000$, using a nominally two-dimensional airfoil with sinusoidal tubercles having amplitude and spacing equal to the average values for the humpback whale.

Experimental results obtained by Johari et al. (2007) at $Re = 183{,}000$ also demonstrated disadvantages for NACA63-021 airfoils with tubercles pre-stall. It was found that the stall angle and maximum lift coefficient were reduced and that there was a corresponding drag increase with respect to an unmodified airfoil. However, improvements were noted in the post-stall regime in which airfoils with tubercles achieved lift coefficients as much as 50% greater than the unmodified airfoil (Johari et al. 2007). Based on these results, it was suggested that tubercles could be used as an active control mechanism whereby they would only be deployed in stalled conditions (Johari et al. 2007).

Results obtained experimentally by Hansen et al. (2001) at $Re = 120{,}000$ using NACA0021 and 65-021 airfoils (Fig. 8) showed similar trends to those reported by Stein and Murray (2005) and Johari et al. (2007). More specifically, degradation in lift and drag performance was observed prior to stall and improved lift performance was evident post-stall. Hansen et al. (2011) noted a decrease in the post-stall drag, which was not the case in the results presented by Johari et al. (2007). The post-stall behavior was investigated over a wide range of AoAs from $\alpha = 0°$ to 90° by Zhang et al. (2013) and it was found that tubercles improved the lift-to-drag ratio at all post-stall angles, particularly in the range between $\alpha = 20°$ to 50°. Performance degradation was observed at pre-stall angles, as consistent with previous studies. These researchers used NACA63-021 airfoils at $Re = 50{,}000$.

While the majority of research in this area has focussed on investigating airfoils having a continuous arrangement of tubercles on the LE, Arai et al. (2010) investigated the effects of placing tubercles in selected locations along an otherwise straight LE. Numerical simulations were performed at $Re = 138{,}000$, using a NACA0018 airfoil. Results indicated that for the configurations investigated, the best lift performance pre-stall was obtained using one tubercle near each wing tip. However, the lift curve was very similar to the curve for the equivalent airfoil with a straight LE and

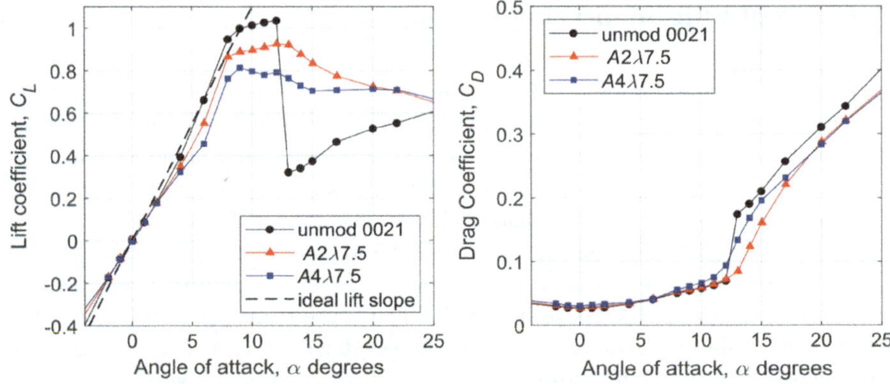

Fig. 8 Lift coefficient (left) and drag coefficient (right) for a NACA 0021 airfoil at $Re = 120{,}000$ (Hansen et al. 2011). Comparison is made between an unmodified and modified airfoils, where $A2\lambda7.5$ ($A/c = 0.03$, $\lambda/c = 0.1$) and $A4\lambda7.5$ ($A/c = 0.06$, $\lambda/c = 0.1$) denote the amplitude, A, and wavelength, λ, in mm of the sinusoidal tubercles

post-stall lift was degraded in a similar manner. A slight increase in both pre-stall and post-stall lift generation was obtained using an airfoil with 4 tubercles in the centre, as opposed to 6 tubercles covering the entire LE. Thus, it was concluded that the tubercles near the wing-tips do not enhance lift performance for a full-span airfoil.

Results from experiments undertaken by de Paula et al. (2017) revealed that tubercles are more beneficial, in terms of performance, for thicker airfoil profiles. These researchers investigated the lift and drag characteristics of NACA0030 airfoils with various tubercle configurations at Reynolds numbers between 50,000 and 290,000. At $Re = 120{,}000$, a significant improvement in performance was shown for the $A/c = 0.03$ and $\lambda/c = 0.11$ tubercle configuration, where c is the mean chord, whereby an increase in 19.4 and 44% of the maximum lift coefficient and stall angle, respectively, were observed. There was also a significant drag reduction both pre-stall and post-stall. Results obtained by de Paula et al. (2017) at higher Reynolds numbers showed degraded pre-stall lift and drag performance and improved post-stall lift performance, as consistent with other studies. At a very high $Re = 1{,}500{,}000$, Rostamzadeh et al. (2017) showed that pre-stall lift was slightly higher, whereas post-stall lift was lower for an airfoil with tubercles compared to one with a straight LE (Fig. 9). On the other hand, stall was more gradual for the tubercle configuration. These results were obtained through numerical simulations using a NACA0021. It was demonstrated by Rostamzadeh et al. (2017) that at $Re = 1{,}500{,}000$, the LSB over the unmodified airfoil does not burst as it does for a lower $Re = 120{,}000$ (Hansen et al. 2011). Therefore, a sudden loss of lift is not observed (Rostamzadeh et al. 2017).

At $Re < 100{,}000$, it was found that the NACA0030 airfoil with a smooth LE demonstrated negative lift characteristics (Bolzon et al. 2015), which is a phenomenon that has been observed in other work (Hansen et al. 2014; Marchaj 1979). This undesirable behavior could be eliminated using certain LE tubercle configurations, resulting in improved lift performance. Serson et al. (2017) also demonstrated numerically that

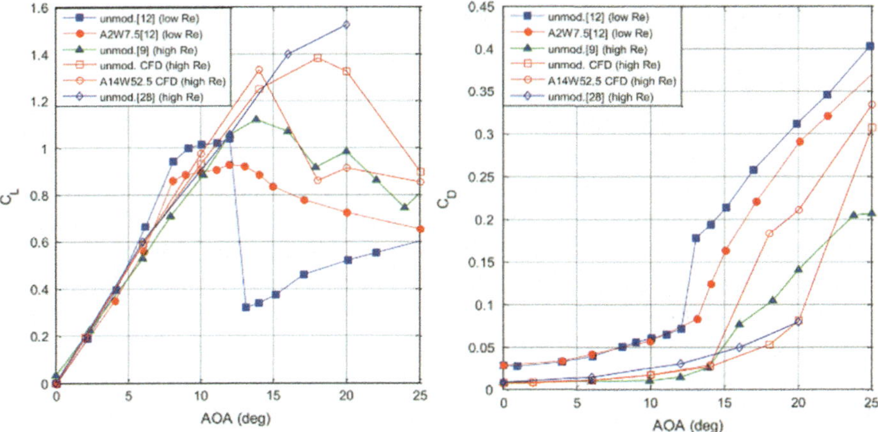

Fig. 9 The lift and drag coefficient curves for the tubercled and unmodified foils at several AoAs [$Re = 120{,}000$ (low) and $Re = 1{,}500{,}000$ (high)] (Rostamzadeh et al. 2017). Here, A2W7.5 ($A/c = 0.03$, $\lambda/c = 0.1$) and A14W52.5 ($A/c = 0.03$, $\lambda/c = 0.1$) denote the amplitude, A, and wavelength, λ, in mm of the sinusoidal tubercles

improved lift and drag performance could be achieved at $Re = 10{,}000$ and $50{,}000$ for a NACA0012 airfoil. It was found that the presence of LE tubercles could either increase or decrease the lift coefficient, depending on the particular Reynolds number and flow regime (Serson et al. 2017). The drag was consistently lower for airfoils with tubercles. Favier et al. (2011) performed direct numerical simulations (DNS) on infinite wings at a post-stall AoA of $\alpha = 20°$ and $Re = 800$. Comparison between results obtained with tubercled and straight LEs indicated a 35% reduction in drag, but there was also a significant reduction in lift.

The effect of tubercles on hysteresis effects was studied experimentally by Cai et al. (2017), through increasing and decreasing the AoA of a NACA63-021. The results showed that the hysteresis effect for airfoils with LE tubercles is very weak, suggesting that the aerodynamic performance is more stable compared to a straight LE. These results are important for airfoils operating in complicated and changeable natural environments (Cai et al. 2017), where there is a tendency for the airfoil to stall regularly.

In summary, for full-span wings, there are negligible pre-stall lift or drag benefits of incorporating tubercles into the LE, except if the wing is very thick or $Re < 100{,}000$. In contrast, the post-stall lift of airfoils with LE tubercles is consistently higher for full-span wings, except at high Reynolds numbers ($>1{,}000{,}000$). Some studies have also reported a corresponding decrease in drag post-stall, although this finding is not unanimous.

5 Finite-Span Unswept Wings

Several studies have focussed on performance differences between finite-span model humpback whale flippers with and without tubercles (Fish and Battle 1995; Miklosovic et al. 2004, 2007; van Nierop et al. 2008; Pedro and Kobayashi 2008; Murray et al. 2005). The majority of these studies reported that model whale flippers with tubercles showed improved lift performance and that there was a negligible effect on the drag (Miklosovic et al. 2004, 2007; Pedro and Kobayashi 2008; Murray et al. 2005). However, reduced pre-stall lift has also been reported (Fish and Battle 1995; van Nierop et al. 2008). Since the humpback whale flipper consists of a complicated geometry with sweep, taper and variable tubercle size and spacing, many researchers have instead chosen to investigate simpler geometries such as an unswept, untapered, finite-span wing in order to focus on the fluid mechanics of the tubercles themselves. The results of this research will be described in the following paragraphs.

An inviscid numerical model of a NACA63-021 airfoil reported by Watts and Fish (2001) showed that incorporation of tubercles into the LE gave rise to a 4.8% increase in maximum lift, 10.9% reduction in induced drag, and 17.6% increase in lift-to-drag ratio at $\alpha = 10°$. Viscous calculations indicated that tubercles have a negligible effect on drag at zero AoA but an 11% increase in form drag was calculated for $\alpha = 10°$. A finite-span (aspect ratio, AR = 2.04) with sinusoidally-shaped tubercles was chosen for this analysis.

Comparison was made between a large range of planform shapes with LE tubercles, including full-span and finite-span models, in an experimental investigation by Custodio et al. (2015). It was found that specific finite-span models with tubercles had higher lift coefficients than an unmodified airfoil. However, the lift-to-drag ratios of all tubercled models were comparable to or less than the equivalent baseline model. The finite-span models had an $AR = 4.3$ and all airfoils followed a $NACA63_4$-021 profile shape. Reynolds numbers between 90,000 and 450,000 were explored. All finite-span models had a rounded tip designed to eliminate the sharp, flat edge at the free tip.

Chen et al. (2012) experimentally investigated the lift and drag performance of $AR = 1$, 2 and 3 NACA0012 airfoils at $Re = 123,000$. It was found that the tubercle configurations tested did not improve the lift performance for any of the finite-span wings, compared to the unmodified airfoil. The maximum lift-to-drag ratio improved in the pre-stall zone for the $AR = 3$ airfoil and relatively large amplitude tubercles. The post-stall drag was lower for all airfoils with tubercles, regardless of the aspect ratio. On the other hand, the post-stall lift was only higher for tubercled airfoils with $AR > 1$.

Experimental results reported by Hansen et al. (2010) at $Re = 120,000$ revealed that the performance of $AR = 3.6$ finite-span airfoils was not enhanced by the presence of LE tubercles. It was shown that the maximum lift coefficient was reduced for finite-span airfoils with tubercles and the pre-stall drag was higher, as shown in Fig. 6. Post-stall, the airfoil with tubercles demonstrated improved performance in terms of both lift and drag, which was consistent with the results obtained using a full-span model.

The performance of $AR = 1.0$ and 1.5 NASA LS(1)-0417 tubercled wings was investigated experimentally at $Re = 70{,}000$ and $140{,}000$ by Guerreiro and Sousa (2012). At the higher Reynolds number, increased lift was observed in the post-stall zone, which was consistent with the results of Hansen et al. (2010) at a similar Reynolds number. At the lower Reynolds number, the lift generated by the wings with tubercles demonstrated higher lift over a larger range of AoA compared to the unmodified wings. This was due to the poor performance of the unmodified airfoil.

Yoon et al. (2011) investigated the effect of varying the proportion of the LE containing tubercles for an $AR = 1.5$ NACA0020 airfoil at a Reynolds number of 1 $\times\ 10^6$. In this study, the number of tubercles was systematically increased from the wing tip. It was shown that the airfoil with the smallest number of tubercles performed similarly to the unmodified airfoil. However, as the number of tubercles increased, the stall angle decreased and thus the maximum lift coefficient also decreased. The post-stall lift was higher for airfoils with tubercles and the airfoil with tubercles along the entire LE generated the most post-stall lift. The drag was equivalent for all modified and unmodified airfoils prior to stall and post-stall, airfoils with LE tubercles generally demonstrated increased drag.

In summary, the pre-stall lift and drag performance of finite-span rectangular wings is not predicted to be improved through incorporation of LE tubercles based on experimental results reported to date. The numerical results presented by Watts and Fish (2001) showed some promise but the reported trends appear to be applicable to inviscid flow only. The post-stall lift of wings with tubercles was consistently improved, with some studies reporting a corresponding reduction in drag. Therefore, the results for finite-span rectangular wings were found to demonstrate the same trends as those for full-span wings. This implies that the benefits of incorporating tubercles may be restricted to swept wings or even to the whale flipper shape, as will be discussed in the following section.

6 Swept Wing Performance

It is worth noting at this stage that the inspiration behind the many investigations into tubercles was the presence of tubercles on the pectoral flippers of humpback whales, as reported in Fish and Battle (1995). The previous sections of this chapter describe the investigation of the mechanisms and benefits of tubercles on 2D wings and on finite-span rectangular wings whose LEs are, in most cases, normal to the oncoming flow. In the present section we will summarise the research into swept and tapered wings with tubercles, in some cases modelling the whale flippers themselves.

The first of the studies on finite-span wings focused on a model humpback whale flipper utilising a NACA0020 profile shape, with and without tubercles, which were tapered and somewhat swept back towards their tips. These results were reported in Miklosovic et al. (2004, 2007) for $Re = 510{,}000$ to $630{,}000$, where tubercles

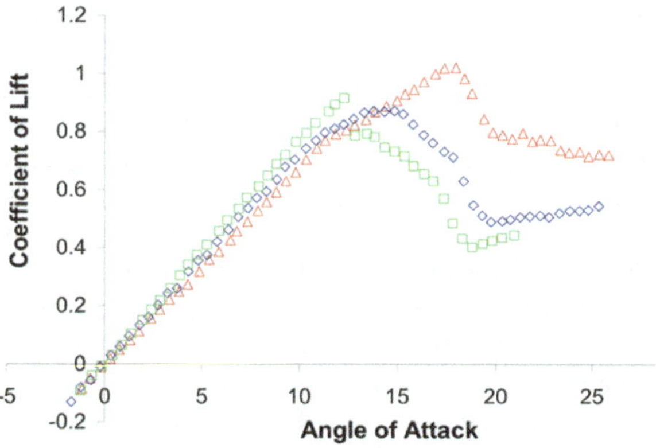

Fig. 10 Comparison of the lift coefficients of a whale flipper model with tubercles for three sweep angles, 0° (green square), 15° (blue diamond) and 30° (red triangle). After Murray et al. (2005)

increased the stall angle by approximately 40% whilst showing minimal change to the lift and drag for AoAs below 9°, a 4–6% higher maximum lift, and improved post-stall performance. As mentioned in the previous section, an additional conclusion of Miklosovic et al. (2007) was that tubercles on flipper models render aerodynamic benefits by inhibiting the spanwise progression of stall. In a related study (Murray et al. 2005), the same team of researchers reported the effects of 15° and 30° sweep using the same flipper-based model. The experiments showed that the addition of tubercles increased the stall angles by 48% and 21% at 15° and 30° sweep respectively, as illustrated in Fig. 10. Increases in maximum lift of 9% and 4% respectively were also found, with similar drag trends to the 0° sweep case. Similar results to these have also been obtained by Wei et al. (2018a) and Kim et al. (2018).

Bolzon et al. (2015, 2016, 2017) reported a series of experiments on a finite-span NACA0021 wing with a taper ratio of 0.4 and sweep of 35° at $Re = 225,000$. Two airfoils were studied: a reference airfoil without tubercles and a modified airfoil with tubercles aligned in the streamwise direction. These researchers combined direct force measurements with wake velocity surveys to explore the relative changes in the induced and profile drag coefficients in detail across the span. At pre-stall up to 8°, Bolzon et al. (2017) found that tubercles reduced the wing's lift coefficient and profile drag coefficient by 4–6% and 7–9.5%, respectively, and increased the lift-to-drag ratio by 2–6%. The key findings of this work were that tubercles spatially modulated the profile drag coefficient across the span, reducing the drag coefficients behind the peaks where attached flow occurs, and increasing the drag coefficient behind the troughs where flow separation is observed. The reverse was found with the induced drag, with induced drag maxima occurring behind the peaks, and minima occurring behind the troughs. The results also revealed that tubercles reduced the drag coefficient near the wing-tip at zero incidence, but at pre-stall AoAs there was little change (Bolzon 2017). It was also noted that the tubercles did not affect the strength of the tip vortex up to an AoA of 3°, but at 6° and above the tubercles reduced the strength of the wing-tip vortex (Bolzon et al. 2016). In general, the change in the profile, induced, and total drag coefficients primarily occurred across the wing-span; there was little contribution from the wing-tip region.

Bolzon et al. (2016) also noted that on the swept wing, the different angles of the sides of each tubercle relative to the oncoming flow, caused one vortex in each pair to become at least four times as strong as its counter-rotating partner. The strengths of the vortices also increased with the AoA. Wei et al. (2018b) compared the effect of tubercles aligned in the streamwise direction with those aligned normal to the LE on a tapered, swept wing at $Re = 82,000$. Wei et al. (2018b) noted that although the differences in alignment did not change the surface flow features significantly, the tubercles aligned normal to the LE led to higher lift-to-drag ratios for AoAs between $2°$ and $7°$. Flow visualization studies showed that each tubercle produces a pattern of nodes and saddles in the surface flow that is significantly tilted due to the presence of the sweep angle. It is noted that the streamwise vortex structures also disrupt the formation of large-scale recirculating regions that are associated with wing-root stall.

Kim et al. (2018) investigated the flow around swept and tapered wings similar to the tubercled and unmodified models used by Miklosovic et al. (2004, 2007), at $Re = 180,000$. In these experiments the delay in the stall and increase in the maximum lift coefficient are consistent with previous studies. The key finding of this study was that, as the AoA increased, longitudinal vortices formed by the tubercles blocked the inboard progression of tip stall compared with the baseline wing model, resulting in delayed stall from $8°$ to $15°$ AoA. It was noted that two vortex pairs formed behind each tubercle: one emanating from the tubercle itself, and another after the mid-chord region. At a post-stall AoA of $16°$, the downwash induced by the tubercles was noted as being responsible for increased lift compared with the unmodified wing.

The results summarised in this section help to explain and confirm the original observations of the effect of LE tubercles on swept and tapered wings by Miklosovic et al. (2007). This discussion represents a small portion of each comprehensive and detailed study, and it is recommended that the reader consult these sources for further study.

7 Aeroacoustics

The use of LE and TE serrations to mitigate noise is relatively well-known and has been the subject of investigation for a considerable period of time (Graham 1934; Soderman 1973; Hersh et al. 1974). One reason for this is that LE and TE serrations have been observed on the plumage of owls, as shown in Fig. 11, which exhibit extremely quiet flight (Bachmann et al. 2007; Sarradj et al. 2010; Wagner et al. 2017). On the other hand, interest in noise reduction through LE tubercles is relatively recent. The application of LE tubercles as a noise reduction mechanism is not immediately obvious in nature. Nevertheless, recent research has revealed that there are clear benefits in using LE tubercles for noise reduction.

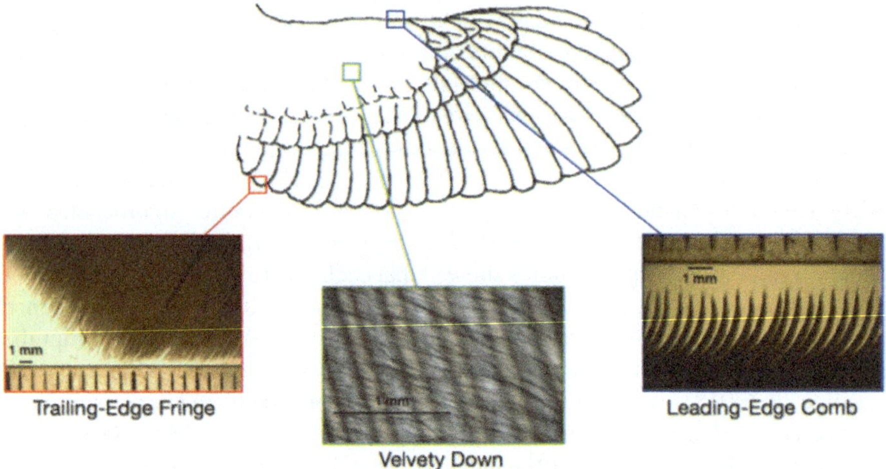

Fig. 11 The three unique wing features believed to make owls fly silently. *Credit* J. W. Jaworski, I. Clark

Substantial reduction of tonal airfoil self-noise was observed by Hansen et al. (2011) using LE tubercles on a NACA0021 airfoil at $Re = 120{,}000$. This type of noise has been identified as a potential problem for wind turbines, gliders, small aircraft, rotors and fans (McAlpine et al. 1999; Kingan and Pearse 2009). Tonal noise also occurs in underwater applications such as hydrofoils, propellers, fast yachts and dinghys (McAlpine et al. 1999). Tonal noise generation is believed to be initiated by Tollmein-Schlichting instabilities in a laminar boundary layer (Custodio 2007; Bolzon et al. 2016; Hansen et al. 2016), which become amplified at the airfoil TE or at a point nearby (McAlpine et al. 1999).

Thus, it was postulated that tonal noise elimination is facilitated by the presence of streamwise vortices generated by the tubercles and that the spanwise variation in separation location is also an important factor (Hansen et al. 2012). Both characteristics modify the boundary layer stability, altering the frequency of velocity fluctuations in the shear layer near the TE. This affects the coherence of the vortex generation downstream of the TE, hence leading to a decrease in TE noise generation as discussed by Hersh et al. (1974). An additional effect is the confinement of the suction surface separation bubble to the troughs between tubercles, which may reduce the boundary layer receptivity to external acoustic excitation. Zhang and Frendi (2016) observed that LE tubercles break up the large vortical structures into smaller ones. These observations were based on the results of an unsteady hybrid Reynolds-Averaged Navier-Stokes and Large Eddy Simulation (RANS-LES) at $Re = 480{,}000$. Noise reductions between 4 and 10 dB were obtained in this study.

Hansen et al. (2011) carried out an experimental investigation at $Re = 120{,}000$ in an anechoic chamber. Through systematic variation of the amplitude and wavelength of the tubercles, it was found that the most successful tubercle configurations for removing tonal noise are those with larger values of A/λ ratio, as shown in Fig. 12b–d,

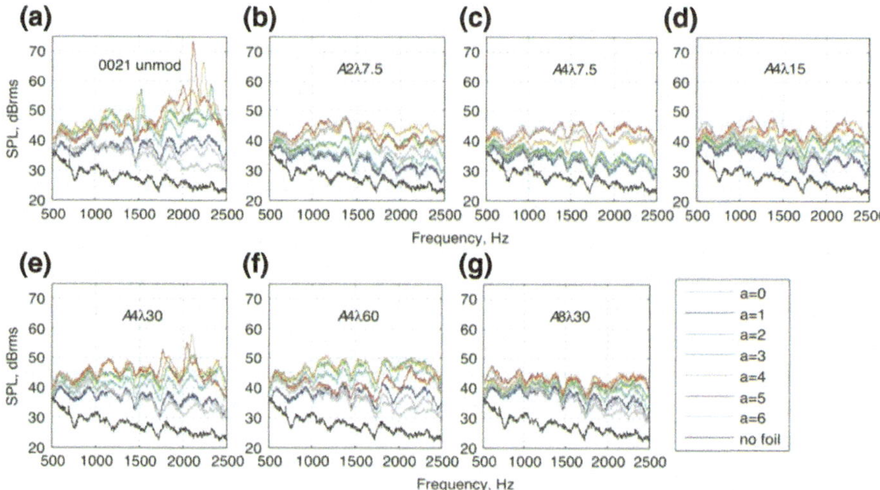

Fig. 12 SPL against frequency measured in an anechoic wind tunnel (AWT) for **a** unmodified 0021, **b** $A2\lambda7.5$, **c** $A4\lambda7.5$, **d** $A4\lambda15$, **e** $A4\lambda30$, **f** $A4\lambda60$, **g** $A8\lambda30$, $Re = 120{,}000$ (Hansen et al. 2012)

g. The noise reductions could be higher than 20 dB and included elimination of tonal noise as well as the associated broadband hump as shown in Fig. 12. Similar acoustic results were also observed by Chong et al. (2015), who showed that the largest amplitude and smallest wavelength tubercles were most effective in reducing tonal noise for a NACA65-(12)10 airfoil at $Re = 150{,}000$. While tonal noise was removed, a decline in lift and drag performance was noted at pre-stall AoAs (Hansen et al. 2011; Chong et al. 2015) post-stall, lift and drag performance was generally improved by the presence of the tubercles (Hansen et al. 2011), however for large-amplitude tubercles, post-stall lift was reduced (Chong et al. 2015).

LE tubercles have also been identified as an effective method for reducing broadband noise that is generated due to turbulent in-flow (Biedermann et al. 2017). This type of noise is also referred to as "airfoil-turbulence interaction (ATI)" noise and is a major contributor to the total noise radiated by rotating machinery. For instance, many fans operate in highly disturbed flows as a consequence of installation parameters that cannot be controlled. ATI is also a significant noise source for wind turbines, propellers, aero-engines and high-lift devices (Kim et al. 2016).

Several researchers have established that LE tubercles can effectively reduce the magnitude of ATI noise using experimental, numerical and analytical models of both airfoils and flat plates. Chaitanya et al. (2017) observed that the noise reduction spectra for airfoils closely follows the behavior observed for flat plates. Moreover, use of flat plates reduces the generation of TE noise, which can be of the same order of magnitude as ATI noise (Roger 2016). A detailed comparison between noise reductions for flat plates and airfoils with tubercles is provided in Chaitanya et al. (2015). Researchers in this area are inclined to use the terminology "LE serrations" as the tubercle amplitudes are relatively large, resulting in sharp peaks. However, for consistency, we will refer to the LE geometry as tubercles here.

The mechanisms that have been identified as being responsible for ATI noise reduction include reduced pressure fluctuations in the "hill" regions (midpoint between the tubercle peaks and troughs) and phase interference effects between the surface pressure fluctuations at the peak and hill regions of the LE geometry, as evident in Fig. 13 (Kim et al. 2016; Lau et al. 2013). The result of these effects is a reduction in the levels of radiated sound pressure, with a corresponding noise reduction in the far field. Reductions in the overall sound pressure level (OASPL) of up to 7 dB have been achieved. Over a specific frequency range of interest, the noise reduction has been measured as greater than 10 dB. Moreover, as the turbulence intensity of the freestream flow increases, the relative reduction in noise associated with LE tubercles is also expected to increase (Biedermann et al. 2017), which is a useful property.

Several tubercle parameters have been found to influence the effectiveness of broadband noise reduction, which include Reynolds number, turbulence intensity, tubercle A/λ ratio and AoA. Using a statistical-empirical model based on experiments, Biedermann et al. (2017) found that the tubercle amplitude was the most significant parameter responsible for variations in the OASPL. However, while increasing the tubercle amplitude was found to reduce the OASPL, this has also been shown to reduce lift and increase drag (Chaitanya et al. 2017). An almost linear relationship between ΔOASPL and amplitude is evident until a certain point, at which an asymptotic level is reached as the amplitude is increased further, as shown in Fig. 14. This is

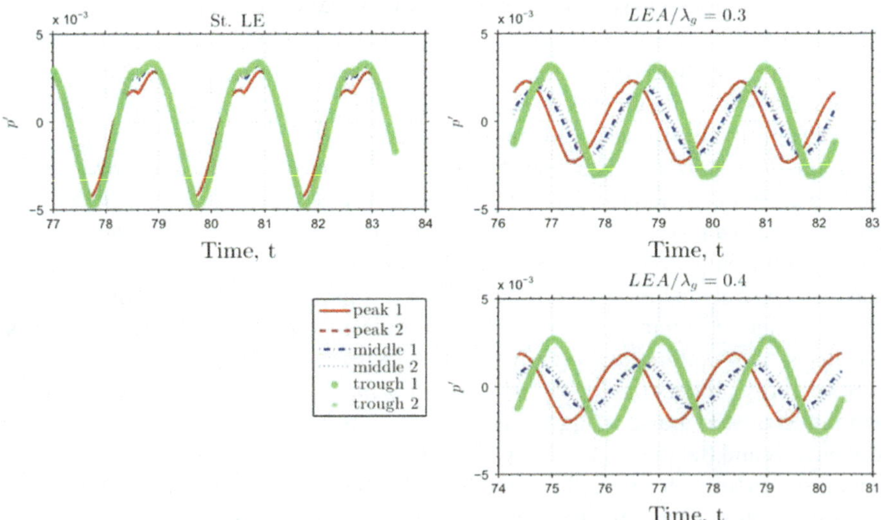

Fig. 13 Time signals of the pressure fluctuations, measured at 6 transducer locations for three different LE geometries. Top left—straight LE. Top and bottom right—LE amplitude (LEA)/gust wavelength ratio = 0.4 and 0.3, respectively (Lau et al. 2013)

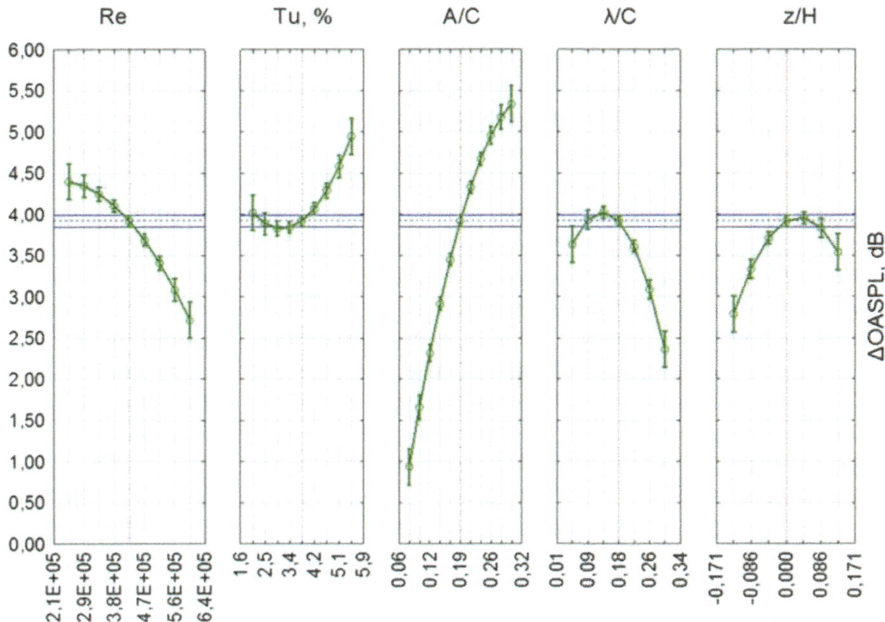

Fig. 14 Impact of parameters that influence the ΔOASPL, including error bounds. The horizontal blue band indicates average noise reduction by the use of serrated LEs (Biedermann et al. 2017)

consistent with results presented by Lau et al. (2013) which showed that the amount of noise reduction was strongly influenced by the ratio between the LE amplitude and the longitudinal wavelength of incident gusts. Turner and Kim (2017) showed that an increase in the tubercle amplitude results in a reduction in source strength at the peak region, until a saturation point is reached. At the trough, the tubercle amplitude has a negligible effect on the source strength. This behavior is attributed to a system of horseshoe-like vortices that are created at the peak region upon impingement of a two-dimensional spanwise vortex structure (Turner and Kim 2017). Results of numerical simulations (Lau et al. 2013) indicated that the maximum noise reduction could be achieved when the ratio between the LE amplitude and the longitudinal gust wavelength exceeds approximately 0.3.

Biedermann et al. (2017) observed that reduction in OASPL generally increased with decreasing wavelength, but that maximum noise reduction occurred at an optimum value, which was not the lowest wavelength investigated. Also, the optimum wavelength was found to be dependent on the turbulence intensity of the incoming flow. The reason for this is that the noise reduction efficiency depends on the relative size of the tubercle wavelength and turbulent eddies. Chaitanya et al. (2017) identified that the optimum wavelength corresponded to when the transverse integral length scale is approximately one-fourth the tubercle wavelength. An interdependence between the airfoil AoA and the tubercle wavelength was also identified by Biedermann et al. (2017).

The reduction in ATI noise that can be achieved using LE tubercles is greatest over specific frequency range (Narayanan et al. 2015; Biedermann et al. 2017). Chaitanya et al. (2017) proposed that an optimum tubercle wavelength can be selected for any arbitrary frequency range and integral length scale as discussed above. At this optimum wavelength, the noise reductions become significant (>3 dB) at a Strouhal number, $St_h = f_{3dB}h/U = 0.25$, where h is half the peak-to-peak amplitude of the tubercles. An inverse relationship was also identified between the Strouhal number and the sound power produced by a wavy LE for $St_h \geq 0.2$ (Chaitanya et al. 2017). As the frequency increases above f_{3dB}, noise reductions generally increase, until the frequency, $f \sim 8U/c_0$, at which point airfoil TE self-noise dominates the interaction noise (Chaitanya et al. 2017). In this high frequency range, the noise is sensitive to the tubercle amplitude, which suggests that LE tubercles can also reduce airfoil self-noise. In addition, LE tubercles could also be incorporated together with TE serrations, in a similar configuration that has already been observed for owl wings (Graham 1934).

In summary, it has been shown that the presence of LE tubercles can result in significant airfoil noise reductions. Noise reductions of as much as 20 dB were obtained for tonal self-noise, due to the elimination of discrete tonal peaks. Tubercle configurations with the largest amplitude-to-wavelength ratios most successfully eliminated this type of noise. It should be noted that a simpler alternative to eliminate this type of tonal noise is through the use of trip strips. Incorporating tubercles into the LE of an airfoil has also resulted in reduction of ATI noise by up to 10 dB. It was revealed that improved noise reduction could be achieved through increasing the tubercle amplitude until a certain point where further amplitude increase did not affect noise reduction. The tubercle wavelength is also important, however, as the optimum length is governed by the length scales dominating the inflow turbulence.

8 Conclusion

A total of nine mechanisms were discussed in an effort to shed light on "how tubercles work", and it is clear that many of these mechanisms are closely related. For example, the vortex generator and delta wing analogies, the observed vorticity canopy and the variation in effective AoA are all in agreement with the observations of streamwise vortices being formed behind the tubercles. This in turn is supported by the variation in wing loading which is linked to the vorticity distribution through Prandtl's lifting-line theory. Some of these observations may also be influenced by the Reynolds number of the flow in which they were observed, where many observations show the complexity of the flow patterns and separation of the various surface flow features is highly Reynolds number dependent. Nevertheless, the broad consensus of these numerous studies is consistent with the findings of Miklosovic et al. (2007), namely that tubercles generate regions of separated and attached flow (similar to confined stall cells located in the troughs) at AoAs below the stall angle of the unmodified wing, and then maintain this pattern until deep stall occurs at a substantially higher

AoA. This explains the decreased maximum lift coefficient in the 2D wing case, as the peak suction is reduced in the separation zones. However, in the 3D swept case, this effect is outweighed by a reduction in spanwise stall progression, which yields a higher maximum lift coefficient. As a result of the partially-attached boundary layer after stall onset, the drag remains lower than that of the unmodified wing until the wing is fully stalled. Thus, whilst the pre-stall performance for unswept wings is generally degraded, stall is more gradual and the post-stall performance is substantially improved. The presence of the axial vortices also leads to a substantial disruption in the generation of tonal noise produced at the TE at Reynolds numbers in the transitional range.

In the case of unswept finite-span (rectangular) wings, results to date show that tubercles generally degrade the pre-stall performance (*lift*, *drag* and lift-to-drag ratio) of the wing, but consistently improve the post-stall lift, with some studies showing a corresponding post-stall drag reduction. Observations suggest that the unswept finite-span mechanism is much the same as for unswept full-span wings, but it is found that for the finite-span swept wing case, the additional inward progression of separated flow from the tip is interrupted by the pattern of streamwise vortices behind the tubercles, as first pointed out by Miklosovic et al. (2007). This aids in maintaining substantially-attached flow to a higher AoA, hence producing a larger maximum lift coefficient without any substantial change in drag.

The effects of tubercles on drag are less easily understood, but it is clear that the presence of tubercles modulates the profile drag across the span, with increased drag behind the troughs and decreased drag behind the peaks. The presence of the tubercles also causes a spanwise-cyclic fluctuation in the lift distribution, which in turn affects the strengths of the streamwise vortices, hence the induced drag distribution. At pre-stall AoAs above $6°$, the tubercles were found to reduce the strength of the wingtip vortex.

To date, there has been no comprehensive experimental study to investigate the effect of Reynolds number on the efficacy of tubercle modifications. One reason for this is the lack of availability of large wind tunnels for such research, with the preponderance of experimental work being limited to Reynolds numbers in the range of $40,000 < Re < 500,000$. Nevertheless, some computational studies have been performed at high Reynolds numbers and these support the notion that tubercles are most effective in the transitional range, where they have greatest impact on the stall process, and will be less effective in the fully-turbulent regime ($Re > 1,000,000$).

References

Arai H, Doi Y, Nakashima T, Mutsuda H (2010) A study on stall delay by various wavy LEs. J Aero Aqua Bio-mech 1(1):18–23

Bachmann T, Klän S, Baumgartner W, Klaas M, Schröder W, Wagner H (2007) Morphometric characterisation of wing feathers of the barn owl Tyto alba pratincola and the pigeon Columba livia. Front Zool 4(1):23

Biedermann TM, Chong TP, Kameier F, Paschereit CO (2017) Statistical–empirical modeling of airfoil noise subjected to LE serrations. AIAA J 55(9):3128–3142

Bolzon MD (2017) The effects of tubercles on swept wing performance at pre-stall angles of attack. Ph.D., The University of Adelaide

Bolzon MD, Kelso RM, Arjomandi M (2015) Tubercles and their applications. J Aerosp Eng 29(1):04015013

Bolzon MD, Kelso RM, Arjomandi M (2016) Formation of vortices on a tubercled wing, and their effects on drag. Aerosp Sci Technol 56:46–55

Bolzon MD, Kelso RM, Arjomandi M (2017) Force measurements and wake surveys of a swept tubercled wing. J Aerosp Eng 30(3):04016085

Cai C, Zuo Z, Liu S, Wu Y (2015) Numerical investigations of hydrodynamic performance of hydrofoils with LE protuberances. Adv Mech Eng 7(7):1687814015592088

Cai C, Zuo Z, Maeda T, Kamada Y, Li Q, Shimamoto K, Liu S (2017) Periodic and aperiodic flow patterns around an airfoil with LE protuberances. Phys Fluids 29(11):115110

Camara JFD, Sousa JMM (2013) Numerical study on the use of a sinusoidal LE for passive stall control at low Reynolds number. In: Proceedings of 51st AIAA Aerospace Sciences Meeting, Grapevine, Texas, AIAA-2013-0062

Chaitanya P, Narayanan S, Joseph P, Vanderwel C, Kim JW, Ganapathisubramani B (2015) Broadband noise reduction through LE serrations on realistic aerofoils. In: Proceedings of 21st AIAA/CEAS aeroacoustics conference, vol 2202

Chaitanya P, Joseph P, Narayanan S, Vanderwel C, Turner J, Kim JW, Ganapathisubramani B (2017) Performance and mechanism of sinusoidal LE serrations for the reduction of turbulence–aerofoil interaction noise. J Fluid Mech 818:435–464

Chen J, Li S, Nguyen V (2012) The effect of LE protuberances on the performance of small aspect ratio foils. In: Proceedings of 15th international symposium on flow visualization, pp 25–28

Chong TP, Vathylakis A, McEwen A, Kemsley F, Muhammad C, Siddiqi S (2015) Aeroacoustic and aerodynamic performances of an aerofoil subjected to sinusoidal LEs. In: Proceedings of 21st AIAA/CEAS aeroacoustics conference, AIAA 2015-2200

Custodio D (2007) The effect of humpback whale-like LE protuberances on hydrofoil performance. Master of Science Thesis, Worcester Polytechnic Institute

Custodio D, Henoch CW, Johari H (2015) Aerodynamic characteristics of finite span wings with LE protuberances. AIAA J 53(7):1878–1893

de Paula AA (2016) The airfoil thickness effects on wavy LE phenomena at low Reynolds number regime. Ph.D., University of Sao Paulo

de Paula AA, Meneghini JK, Kleine VG, Girard RM (2017) The wavy LE performance for a very thick airfoil. In: Proceedings of 55th AIAA Aerospace Sciences Meeting, Texas, USA, AIAA2017-0492

Dropkin A, Custodio D, Henoch CW, Johari H (2012) Computation of flow field around an airfoil with LE protuberances. J Aircraft 49(5):1345–1355

Favier J, Pinelli A, Piomelli U (2011) Control of the separated flow around an airfoil using a wavy LE inspired by humpback whale flippers. CR Mec 3401(1):107–114

Fish FE, Battle JM (1995) Hydrodynamic design of the humpback whale flipper. J Morphol 225(1):51–60

Graham R (1934) The silent flight of owls. Aeronaut J 38(286):837–843

Guerreiro JE, Sousa JM (2012) Low-Reynolds-number effects in passive stall control using sinusoidal LEs. AIAA J 50(2):461–469

Hansen KL, Kelso RM, Dally BB (2010) An investigation of three dimensional effects on the performance of tubercles at low Reynolds numbers. In: Proceedings of 17th Australian fluid mechanics conference, Auckland, New Zealand

Hansen KL, Kelso RM, Dally BB (2011) Performance variations of LE tubercles for distinct airfoil profiles. AIAA J 49(1):185–194

Hansen KL, Kelso RM, Doolan C (2012) Reduction of flow induced tonal noise using LE sinusoidal modifications. Acoust. Aust. 40(3)

Hansen K, Kelso R, Choudhry A, Arjomandi M (2014) Laminar separation bubble effect on the lift curve slope of an airfoil. In: Proceedings of 19th Australasian fluid mechanics conference. Melbourne, Australia

Hansen KL, Rostamzadeh N, Kelso RM, Dally BB (2016) Evolution of the streamwise vortices generated between LE tubercles. J Fluid Mech 788:730–766

Hersh AS, Sodermant PT, Hayden RE (1974) Investigation of acoustic effects of LE serrations on airfoils. J Aircraft 11(4):197–202

Johari H, Henoch C, Custodio D, Levshin A (2007) Effects of LE protuberances on airfoil performance. AIAA J 45(11):2634–2642

Kim JW, Haeri S, Joseph PF (2016) On the reduction of aerofoil–turbulence interaction noise associated with wavy LEs. J Fluid Mech 792:526–552

Kim H, Kim J, Choi H (2018) Flow structure modifications by LE tubercles on a 3D wing. Bioinspir Biomim 13(6):066011

Kingan MJ, Pearse JR (2009) Laminar boundary layer instability noise produced by an aerofoil. J Sound Vib 322(4–5):808–828

Lau ASH, Haeri S, Kim JW (2013) The effect of wavy LEs on aerofoil–gust interaction noise. J Sound Vib 332:6234–6253

Lin JC (2002) Review of research on low-profile vortex generators to control boundary layer separation. Prog Aerosp Sci 38(4–5):389–420

Malipeddi AK, Mahmoudnejad N, Huffmann KA (2012) Numerical analysis of effects of LE protuberances on aircraft wing performance. J Aircraft 49(5):1336–1344

Marchaj CA (1979) Aero-hydrodynamics of sailing. Adlard Coles Ltd/Granada

McAlpine A, Nash E, Lowson M (1999) On the generation of discrete frequency tones by the flow around an aerofoil. J Sound Vib 222(5):753–779

Miklosovic DS, Murray MM, Howle LE, Fish FE (2004) LE tubercles delay stall on humpback whale flippers. Phys Fluids 16(5):L39–L42

Miklosovic DS, Murray MM, Howle L (2007) Experimental evaluation of sinusoidal LEs. J Aircraft 44(4):1404–1408

Murray MM, Miklosovic DS, Fish FE, Howle LE (2005) Effects of LE tubercles on a representative whale flipper model at various sweep angles. In: Proceedings of 14th unmanned untethered submersible technology (UUST), Durham, New Hampshire, Aug 2005

Narayanan S, Chaitanya P, Haeri S, Joseph P, Kim J, Polacsek C (2015) Airfoil noise reductions through LE serrations. Phys Fluids 27(2):025109

Pedro HTC, Kobayashi MH (2008) Numerical study of stall delay on humpback whale flippers. In: Proceedings of 46th AIAA Aerospace Sciences Meeting and Exhibit, AIAA2008-0584

Perez-Torro R, Kim JW (2017) A large-eddy simulation on a deep-stalled aerofoil with a wavy LE. J Fluid Mech 813:23–52

Roger M (2016) Airfoil turbulence-impingement noise reduction by porosity or wavy LE cut: experimental investigations. In: Proceedings of INTER-NOISE and NOISE-CON Congress and conference, vol 253, No. 2, pp 6366–6375. Institute of Noise Control Engineering

Rostamzadeh N, Kelso RM, Dally BB, Hansen KL (2013) The effect of undulating LE modifications on NACA 0021 airfoil characteristics. Phys Fluids 25(11):117101

Rostamzadeh N, Hansen KL, Kelso RM, Dally BB (2014) The formation mechanism and impact of streamwise vortices on NACA 0021 airfoil's performance with undulating LE modification. Phys Fluids 26(10):107101

Rostamzadeh N, Kelso RM, Dally BB (2017) A numerical investigation into the effects of Reynolds number on the flow mechanism induced by a tubercled LE. Theoret Comput Fluid Dyn 31(1):1–32

Sarradj E, Fritzschey C, Geyery T (2010) Silent owl flight: bird flyover noise measurements. In: Proceedings of 16th AIAA CEAS aeroacoustics conference

Serson D, Meneghini JR (2015) Numerical study of wings with wavy leading and TEs. Procedia IUTAM 14:563–569

Serson D, Meneghini JR, Sherwin JS (2017) Direct numerical simulations of the flow around wings with spanwise waviness. J Fluid Mech 826:714–731

Skillen A, Revell A, Pinelli A, Piomelli U, Favier J (2014) Flow over a wing with LE undulations. AIAA J 53(2):464–472

Soderman PT (1973) LE serrations which reduce the noise of low-speed rotors. NASA Technical Note, NASA TN D-7371

Stanway MJ (2008) Hydrodynamic effects of LE tubercles on control surfaces and in flapping foil propulsion. Master of Science in Ocean Engineering Mechanical Engineering, Massachusetts Institute of Technology, Feb 2008

Stein B, Murray MM (2005) Stall mechanism analysis of humpback whale flipper models. In: Proceedings of unmanned untethered submersible technology (UUST), Durham, NH, Aug 2005

Turner JM, Kim JW (2017) Aeroacoustic source mechanisms of a wavy LE undergoing vortical disturbances. J Fluid Mech 811:582–611

van Nierop E, Alben S, Brenner MP (2008) How bumps on whale flippers delay stall: an aerodynamic model. Phys Rev Lett 100(5):054502

Wagner H, Weger M, Klaas M, Schröder W (2017) Features of owl wings that promote silent flight. Interface Focus 7(1):20160078

Watts P, Fish PE (2001) The influence of passive LE tubercles on wing performance. In: Proceedings of 12th international symposium on unmanned untethered submersible technology (UUST). Autonomous Undersea Systems Institute, Durham, NH

Weber PW, Howle L, Murray M (2010) Lift, drag and cavitation onset on rudders with LE tubercles. Marine Technol 47(1):27–36

Weber P, Howle L, Murray M, Miklosovic D (2011) Computational evaluation of the performance of lifting surfaces with LE protuberances. J Aircraft 48(2):591–600

Wei Z, Lian L, Zhong Y (2018a) Enhancing the hydrodynamic performance of a tapered swept-back wing through LE tubercles. Exp Fluids 59:103

Wei Z, New TH, Cui YD (2018) Aerodynamic performance and surface flow structures of LE tubercled tapered swept-back wings. AIAA J 56(1)

Yoon HS, Hung PA, Jung JH, Kim MC (2011) Effect of the wavy LE on hydrodynamic characteristics for flow around low aspect ratio wing. Comput Fluids 49:276–289

Zhang M, Frendi A (2016) Effect of airfoil LE waviness on flow structures and noise. Int J Numer Meth Heat Fluid Flow 26(6):1821–1842

Zhang M, Wang G, Xu J (2013) Aerodynamic control of low-Reynolds-number airfoil with LE protuberances. AIAA J 51(8):1960–1971

Zhang MM, Wang GF, Xu JZ (2014) Experimental study of flow separation control on a low-Re airfoil using LE protuberance method. Exp Fluids 55(4):1710

Zhao M, Zhang M, Xu J (2017) Numerical simulation of flow characteristics behind the aerodynamic performances on an airfoil with LE protuberances. Eng Appl Comput Fluid Mech 11(1):193–209

Tubercle Geometric Configurations: Optimization and Alternatives

R. Kelso, N. Rostamzadeh and K. Hansen

Abstract This chapter describes efforts by researchers to optimize tubercle configurations and investigate alternative geometries. It begins with a discussion of the optimization of sinusoidal leading-edge tubercles on full-span (2D) foils, focusing mainly on the wavelength (λ) and amplitude (A) relative to the wing's mean chord (c). This is followed by a description of several optimization efforts for finite wings and fan blades, focusing mainly on the arrangement of the tubercles. These include the use of computational algorithms to determine the optimum design. Lastly, the chapter describes several alternative tubercle designs, such as serrated and wavy designs, and alternative versions of tubercles such as dome-shaped protrusions.

Keywords Optimisation · Amplitude · Wavelength · Phase · Alignment · Distribution · Serrated · Wavy · Rippled

1 Optimization of the Tubercle Geometry on Full-Span Airfoils

This section discusses the effects of the wavelength and amplitude of leading-edge (LE) tubercles on full-span foils. Several groups have contributed to the search for an optimum configuration, and some consensus has been achieved, but the data are still somewhat sparse. The different requirements for pre-stall and post-stall have begun to be identified, but the effects of Reynolds number have not been comprehensively explored. It is important to note that the following discussion does not specifically differentiate between the way the tubercles themselves have been designed—e.g. by extending the peaks and troughs as far as the maximum thickness point (see investigations by authors such as Custodio, Chong and others) on the wing, or by extending them progressively to the trailing edge (see investigations by authors such

R. Kelso (✉) · N. Rostamzadeh
School of Mechanical Engineering, University of Adelaide, Adelaide 5005, Australia
e-mail: richard.kelso@adelaide.edu.au

K. Hansen
College of Science and Engineering, Flinders University, Adelaide 5042, Australia

as Murray, Hansen, Rostamzadeh, Bolzon and others). As this design detail produces relatively minor differences in the wing profiles, the tubercle geometries have been treated as being equivalent in the following discussion.

An experimental investigation on the influence of tubercle amplitude and wavelength on airfoil performance was carried out by Johari et al. (2007) using NACA 63-021 airfoils at $Re = 183,000$. The experimental results shown in Fig. 1 indicate that airfoils with smaller amplitude tubercles ($A/c = 0.025$) performed best in terms of stall angle, maximum lift coefficient and reduced drag. On the other hand, larger amplitude tubercles promoted softer stall characteristics ($A/c = 0.12$). Although, the authors commented that the effects of wavelength variation were minor, inspection of the figures presented in their paper reveals that the tubercle configurations with smaller wavelength ($\lambda/c = 0.25$) achieved higher maximum lift coefficient and stall angle with reduced drag. For all tubercle configurations investigated, the post-stall lift-to-drag ratio increased. However, there was no clear relationship between tubercle amplitude and wavelength and improved post-stall performance.

Improved performance for airfoils with relatively small amplitude and wavelength tubercles was also observed in an experimental study conducted by Hansen et al. (2011). This research investigated the performance of NACA 0021 and NACA 65-021 airfoils with tubercles at $Re = 120,000$. An example of the results is shown in Fig. 2. It was found that the highest stall angle, highest maximum lift coefficient and lowest drag could be achieved with the smallest amplitude and wavelength tubercle configuration ($A/c = 0.029$, $\lambda/c = 0.11$), consistent with Johari et al. (2007). It was also observed that the post-stall lift and drag performance for this tubercle configuration was equivalent to or better than that obtained using tubercles with double the amplitude and the same wavelength, in contrast to Johari et al. (2007). Comparison between four different airfoils with the same tubercle wavelength revealed that the there is a certain point at which further reduction in wavelength has a negative

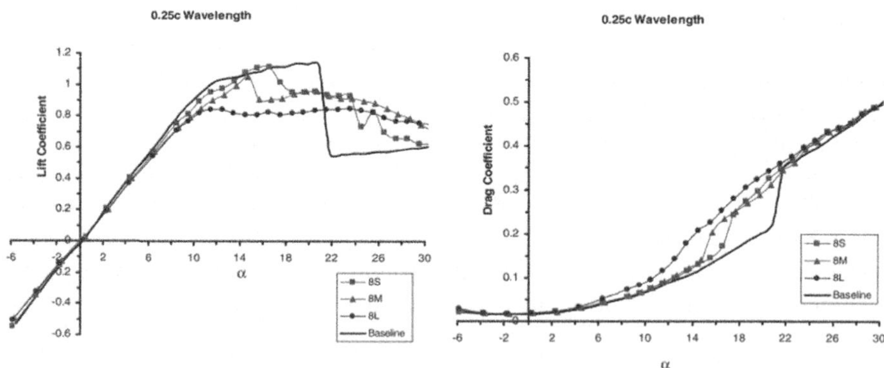

Fig. 1 Lift coefficient (left) and drag coefficient (right) for a NACA 63-021 airfoil at $Re = 183,000$ (Johari et al. 2007). Comparison is made between an unmodified and modified airfoils, where S, M and L (small, medium, large) denote cases $A/c = 0.025, 0.05, 0.12$ respectively and the wavelength is fixed at $\lambda/c = 0.25$

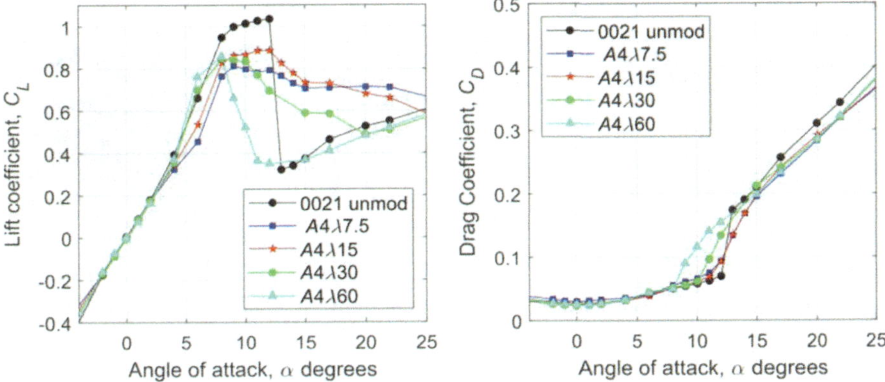

Fig. 2 Lift coefficient (left) and drag coefficient (right) for a NACA 0021 airfoil at $Re = 120{,}000$ (Hansen et al. 2011). Comparison is made between an unmodified and modified airfoil, where A denotes the amplitude and λ the wavelength, in mm, of the sinusoidal tubercles. The amplitude is varied ($A/c = 0.029$, 0.057 and 0.11) and the wavelength is varied ($\lambda/c = 0.11$, 0.21, 0.43 and 0.86)

impact on lift performance, as shown in Fig. 2. This implied that there exists an optimum wavelength for each amplitude and hence, the amplitude-to-wavelength ratio is important. In this study, the optimized amplitude and wavelength configuration gave rise to pre-stall lift performance that approached the values attained by the unmodified airfoil and improved post-stall performance.

Similar trends were reported by de Paula et al. (2017) who showed that improved lift performance could be achieved at a wide range of Reynolds numbers for a NACA 0030 airfoil using the smallest amplitude and wavelength tubercles ($A/c = 0.03$, $\lambda/c = 0.11$). The Reynolds number in this study ranged from 50,000 to 290,000 and the small-amplitude/small-wavelength configuration consistently achieved the highest maximum lift coefficient, stall angle and post-stall performance. This configuration also demonstrated relatively low drag compared to the other tubercle arrangements. At a Reynolds number of 120,000, the pre-stall lift performance of the small amplitude/wavelength configuration even surpassed that of the unmodified airfoil by a significant amount. At low Reynolds numbers of 50,000 and 80,000, airfoils with small amplitude/wavelength tubercles were also found to generate higher pre- and post-stall lift than the unmodified airfoil.

According to experimental results published by Favier et al. (2012), maximum lift generation was achieved when the tubercle amplitude and wavelength were minimized but there was a corresponding increase in the lift and drag fluctuations. The lift performance did not surpass that of the unmodified airfoil for any of the investigated tubercle configurations. These authors investigated a deep stall condition with an angle of attack (AoA) $\alpha = 20°$ and a low Reynolds number of 800. The results of this parametric study, shown in Fig. 3, demonstrate that maximum drag reduction was achieved with large amplitude tubercles ($A/c = 0.07$) and a mid-range value of

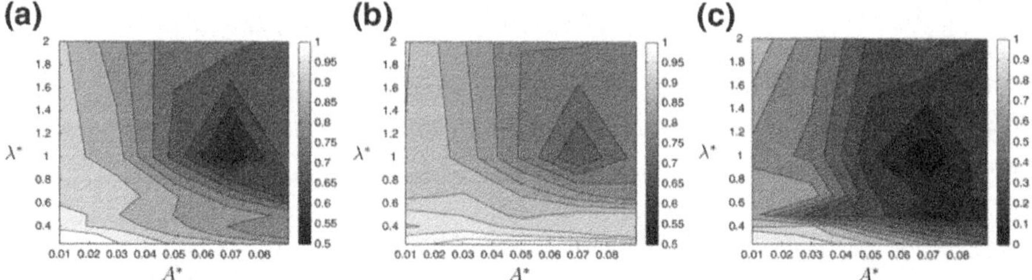

Fig. 3 Contours of performance ratios obtained by varying A/c and λ/c (Favier et al. 2012). **a** R_1 = $<C_D>/<C_{D,baseline}>$, **b** R_1 = $<C_L>/<C_{L,baseline}>$ and **c** R_1 = $<C_L'/<C_{L,baseline}'>$. Improved performance is achieved when $R_1 < 1$, $R_2 > 1$ and $R_3 < 1$

the wavelength ($\lambda/c = 1$). This case was also associated with both a reduction in lift and a reduction in lift fluctuations.

Favier et al. (2012) explained that variations in the tubercle amplitude and wavelength influence the wake topology, the size of the recirculation zone and the strength of the wake vortex shedding. For small amplitude-to-chord ratios, $A/c < 0.07$, the streamwise vortices generated by the tubercles are too weak to stabilize the separation region. For larger amplitude-to-chord ratios, $A/c \geq 0.07$, the frequency spectrum was observed to flatten, indicating a weakening of the wake vortex shedding. The tubercle wavelength was found to influence the separation characteristics, where the flow was fully separated for $A/c = 0.07$ and $\lambda/c = 0.5$ but attached behind the peaks for $A/c = 0.07$ and $\lambda/c = 1$.

In apparent contrast to the results presented above, Biedermann et al. (2017) achieved optimum L/D performance using a mid-range value of the tubercle amplitude ($A/c = 0.19$) and the largest wavelength ($\lambda/c = 0.3$), respectively, of the investigated configurations. However, these results are relevant for an AoA of $10°$ only and results at higher angles were not obtained due to limitations of the experimental apparatus. In this study, a cambered NACA 65(12)-10 airfoil was chosen due to its suitability in rotating-machinery applications and the Reynolds number was 250,000. Acoustic measurements were also recorded and it was observed that there was a lack of correlation between improved aerodynamic performance and increased noise reduction for turbulence-induced noise. In fact, the largest tubercle wavelength demonstrated the worst acoustic performance and noise reduction was found to increase with amplitude.

A parametric experimental study conducted by Chaitanya et al. (2017) found that the small amplitude tubercle configuration ($A/c = 0.13$) achieved the highest lift for all AoA and lowest drag for relatively high AoA compared to the other tubercle configurations investigated. This study focused on the aerodynamic and aeroacoustic performance of a NACA 65(12)10 airfoil with tubercles at small effective AoA and a Reynolds number of 600,000. As observed by Biedermann et al. (2017), optimising the aerodynamic performance came at the expense of degraded acoustic performance (Chaitanya et al. 2017).

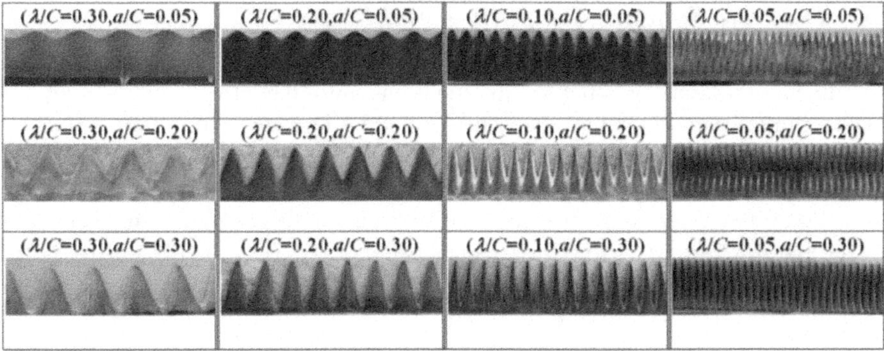

Fig. 4 Various tubercle configurations that were investigated by Chong et al. (2015)

The results of an experimental investigation by Chong et al. (2015) indicated that the smallest amplitude ($A/c = 0.05$) and largest wavelength ($\lambda/c = 0.2$) tubercle configuration produced the best lift performance at all AoA, particularly close to the stall angle and post-stall. The maximum lift coefficient and post-stall performance of this configuration even surpassed that of the unmodified baseline case. Conversely, the large amplitude ($A/c = 0.3$) and small wavelength ($\lambda/c = 0.05$) configuration was shown to be superior for noise reduction purposes. This study investigated the performance of NACA 65(12)10 airfoils with 12 different tubercle configurations, as shown in Fig. 4, at a Reynolds number of 150,000.

2 Optimization of the Tubercle Geometry on Finite-Span Wings

Compared with their full-span counterparts, relatively little work has been done to optimize the tubercle design of finite-span foils. On such foils the optimization process must consider factors such as the tubercle spacing and amplitude relative to the local chord, the alignment of the tubercles relative to the flow stream and the distribution of tubercles along the span. Many experimental optimization studies have been conducted to investigate a limited range of parameters, but in order to expand the range of variables under investigation, a number of studies have been carried out using algorithm-based computational approaches, including the use of genetic algorithms. The following paragraphs describe several of these studies.

2.1 Tubercle Spacing, Amplitude and Phase

Experimentally-based parametric investigations into the effect of tubercles on unswept finite-span wings have been performed by Hansen et al. (2010) and Custodio et al. (2015) as discussed in Chapter "Tubercle Geometric Configurations: Optimization and Alternatives". Hansen et al. (2010) studied NACA 0021 foils at $Re = 120{,}000$, with $AR = 3.6$ and tubercle amplitude ratios (λ/c) ranging from 0.03 to 0.11 and A/λ from 0.07 to 0.53. Custudio et al. (2015a) studied NACA 63_4-021 foils at $Re = 90{,}000$ to 450,000, with $AR = 4.3$ and tubercle amplitude ratios ranging from 0.025 to 0.12 and A/λ from 0.05 to 0.48. Despite the different sectional shapes, tubercle design and Reynolds numbers, both studies demonstrated that foils with smaller amplitude and smaller wavelength tubercles generally performed best in terms of lift, drag and efficiency (or lift-to-drag ratio, L/D). However, neither study conclusively demonstrated an optimum configuration due to the small number of configurations tested.

Several studies have also investigated the aerodynamic effects of tubercles on finite-span wings using numerical algorithms based on classical aerodynamic models and experimental data. Firstly, van Nierop et al. (2008) modelled a high aspect-ratio finite-span foil of elliptical planform using lifting-line theory to account for spanwise variation in circulation due to variations in thickness and chord. The tubercles were modelled as a chord-length perturbation of small amplitude (with a straight trailing-edge (TE)) and a large wavelength relative to the span. The model predicts that the presence of tubercles flattens out the lift curve so that stall is delayed and maximum lift is decreased, and that this effect is insensitive to the tubercle wavelength. It also demonstrates that the maximum lift coefficient occurs when the amplitude approaches zero, i.e. the reference case.

A second study, also based on lifting-line theory was performed by Bolzon et al. (2016) to investigate the effect of the tubercle amplitude, wavelength and phase on the performance of an unswept NACA 0021 wing of aspect ratio 3.5. The study utilized experimental data from Hansen (2012) for a single AoA of 3°, and a Reynolds number of 120,000. The novel aspect of this work was the investigation of the tubercle phase, defined as the point along the wing-tip tubercle at which the wing terminates. The study examined the lift coefficient, induced drag coefficient, and the lift-to-induced-drag ratio, but the profile drag was not considered. The study concluded that the tubercle phase had the greatest effect on wing performance, while the wavelength had the least effect, with the lift-to-induced-drag ratio being increased by as much as 7.7%, corresponding to a wing terminating midway between a tubercle trough and peak. For optimum design, it was recommended that $A/c > 0.21$ and $\lambda/b > 0.5$, where b is the wing span (i.e. a single tubercle or planform similar to a tapered wing). These optimization studies were extended to include a genetic-algorithm-based optimization as reported in Bolzon (2016). Five tubercle parameters were optimized by the algorithm to evolve tubercle geometries that produced the highest lift-to-induced-drag ratio of the wing. These included the amplitude, wavelength, phase, location of the tubercles, and the number of tubercles. The amplitude ranged from $0 < A/c > 0.29$, and the wavelength was limited to less than half of the span to prevent the algorithm from generating monotonically tapered wings. The fitness parameter was chosen to be the lift-to-induced-drag ratio. A total of 400 generations of tubercle configurations were evolved, leading to numerous designs with similar performance, all with a trough immediately inboard of the tip (sometimes referred to as notched wings), and the tip approximately midway between a trough and a peak. The best configuration found using the genetic algorithm increased the lift-to-induced-drag ratio by 4.3% over the equivalent smooth wing. The trends found at an AoA of 3° were also confirmed for lower AoA. Overall, the optimum parameters were $0.25 < A/c > 0.29$, $\lambda/c > 0.94$, with the single tubercle trough beginning at a distance of between 0.27 and 0.58 of the span outboard of the wing root.

In another novel approach to tubercle optimization, Taheri (2018) developed a meta-model using an artificial neural network to investigate and optimize the effects of LE tubercles on the performance of a generalized swept-wing planform. The model was "trained" using the majority of the available experimental data, then coupled with a genetic algorithm to seek an optimal design. The training parameter space defined $0.03 \leq \lambda/c \leq 0.25$ and $0.03 \leq A/c \leq 0.15$, with $5° < \alpha < 40°$ and $Re = 130{,}000$. The wing chosen for optimization had an unswept rectangular planform of aspect ratio 2.5 and a NACA 0021 profile shape. A fitness function was chosen to seek the optimum lift/drag ratio. The optimum lift/drag ratio in this case was found to occur with $A/c \approx 0.03$ and $\lambda/c \approx 0.19$, at an AoA of 28.6°. Investigation using computational fluid dynamics (CFD) showed that this case corresponds to a partially-stalled flow condition.

An optimization study into the effects of tubercle amplitude and spacing has also been carried out in the context of a propeller. Butt and Talha (2018) used CFD to optimise the tubercle amplitude and wavelength on an 8×3.8 (8 inch diameter, 3.8 inch advance per revolution) propeller at a range of rotational velocities and advance ratios. Only two amplitudes and two wavelengths were studied, generating four flow cases. The tubercle amplitudes and wavelengths were specified, but the chord was not stated explicitly. The study found that the configurations with relatively larger tubercle amplitude and smaller tubercle wavelength, i.e. large tubercle amplitude-to-wavelength ratio, were more efficient, with the greatest improvement being 14.2% over the baseline case.

2.2 Tubercle Alignment

One specific tubercle design parameter that has received little attention is the tubercle alignment on swept wings. Whilst the majority of swept-wing studies have used tubercles that are aligned with the direction of the oncoming flow, a study by Wei et al. (2017) also considered tubercles aligned normal to the swept LE, shown as "Modified B" in Fig. 5. One major effect of this variation is that the strengths of the vortices downstream of each side of each tubercle is altered. Wei et al. (2017) found that the tubercles aligned normal to the LE led to higher lift-to-drag ratio for AoAs

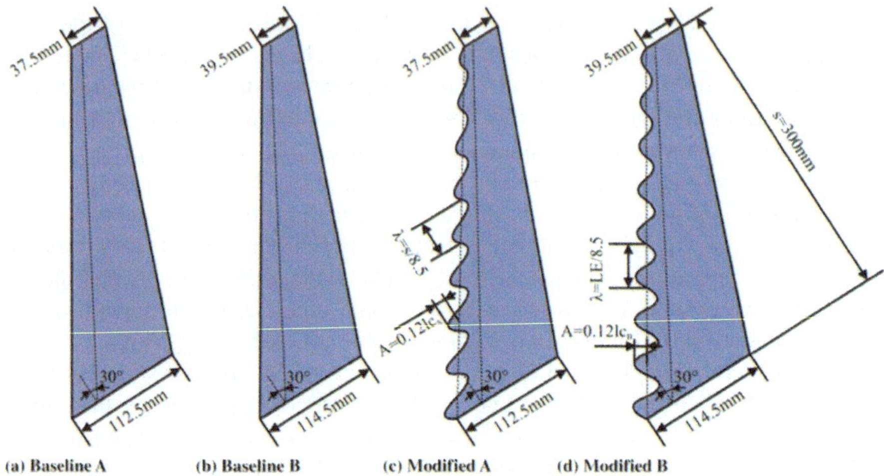

Fig. 5 Geometry of the four tapered swept-back wings used by Wei et al. (2017). Modified A incorporates tubercles aligned with the direction of flow, whilst Modified B has tubercles normal to the flow

between 2° and 7°, although from 7° to 15° the performance was degraded relative to the flow-aligned tubercle design.

2.3 Tubercle Distribution

The above discussion reports a number of studies of foils of various configurations where the tubercles are distributed uniformly along the LE. In only one study did the optimization process lead to a non-uniform distribution, and this was in the case of Bolzon et al. (2016) and Bolzon (2016), where the wavelength of the tubercles exceeded the span of the wing. The following discussion relates to studies where the distribution of tubercles along the span varies, beginning with a generic swept wing, followed by two studies of marine turbines.

The first of these studies was by Yoon et al. (2011), who undertook a numerical study of finite-span wings of aspect ratio 1.5, with tubercles incorporated in a portion of the LE, beginning from the tip. In this numerical investigation where the Reynolds number was maintained at 1,000,000, Rw was defined as the length of the LE covered by tubercles divided by the span of the wing. Five different Rw ratios of 0.2, 0.4, 0.6, 0.8 and 1.0 were studied at a fixed λ/c of 0.2 and $A/c = 0.025$. The study concluded that the finite wing whose Rw was equal to 0.2 (i.e. one tubercle wavelength located adjacent to the wing tip) stalled at 20°, whereas other wings stalled at lower AoAs. In the post-stall regime, the wings with Rw equal to and above 0.4 yielded higher lift coefficients than the unmodified NACA 0020 wing.

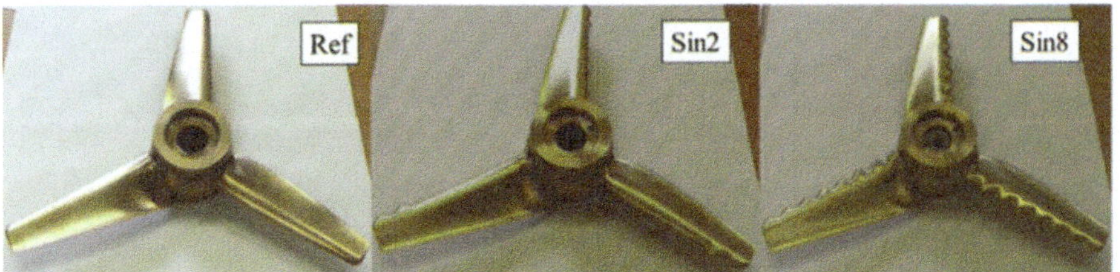

Fig. 6 Marine turbine models investigated by Shi et al. (2017)

The application of tubercles to a marine tidal turbine was first reported by Gruber et al. (2011), who studied three different turbine configurations: a baseline design based on a NACA 633-618 sectional profile, a similar turbine with tubercles on the outer 2/3 of each blade, and one with tubercles on the outer 1/3 of each blade. Here the tubercle A/c was 0.05, with a fixed wavelength so that λ/c varied over a range of approximately 0.3–0.5, depending on spanwise position. Both tubercle configurations demonstrated improved performance compared with the baseline, with the 1/3 tubercle configuration performing consistently better than the others across the full range of velocities tested. At each tip speed ratio, the 1/3 tubercled blades produced a 15–85% higher power coefficient than the control blades. The cut-in velocity of the 1/3 configuration was also consistently lower than measured for the other designs, confirming the improved performance. Later, Shi et al. (2016b) investigated a similar set of marine turbines based on an NREL S814 profile, comparing a reference case with a modified turbine with tubercles across the full span, and one with two tubercles in the outer 25% of the span. Here the tubercle A/c was 0.1, and λ/c varied from 0.3 at the root to 0.6 at the tip. Similar results were achieved to those reported by Gruber et al. (2011), with the best performance again being achieved by the turbine with tubercles near the tip only. Following this work, Shi et al. (2016a, 2017) investigated the mechanism leading to these performance gains using CFD as well as 2D and 3D Particle-image velocimetry, concluding that the main mechanism responsible for enhanced performance was the reduction in the strength of the tip vortex. Examples of the marine turbine geometries are shown in Fig. 6.

2.4 Single Tubercle

Following their earlier work optimising the tubercle distribution, Bolzon et al. (2017) investigated the effects of a single tubercle located at the tip of a swept wing on its performance at pre-stall AoAs. The wing was based on a NACA 0021 profile, had a sweep angle of 35°, and was tested at a Reynolds number of 225,000 based on the mean chord. Three configurations were tested and are described in Fig. 7; one with a smooth LE, one that was smooth from the wing root to 82% of the span, with a single tubercle producing a peak at the wing tip, and a similar design with a single

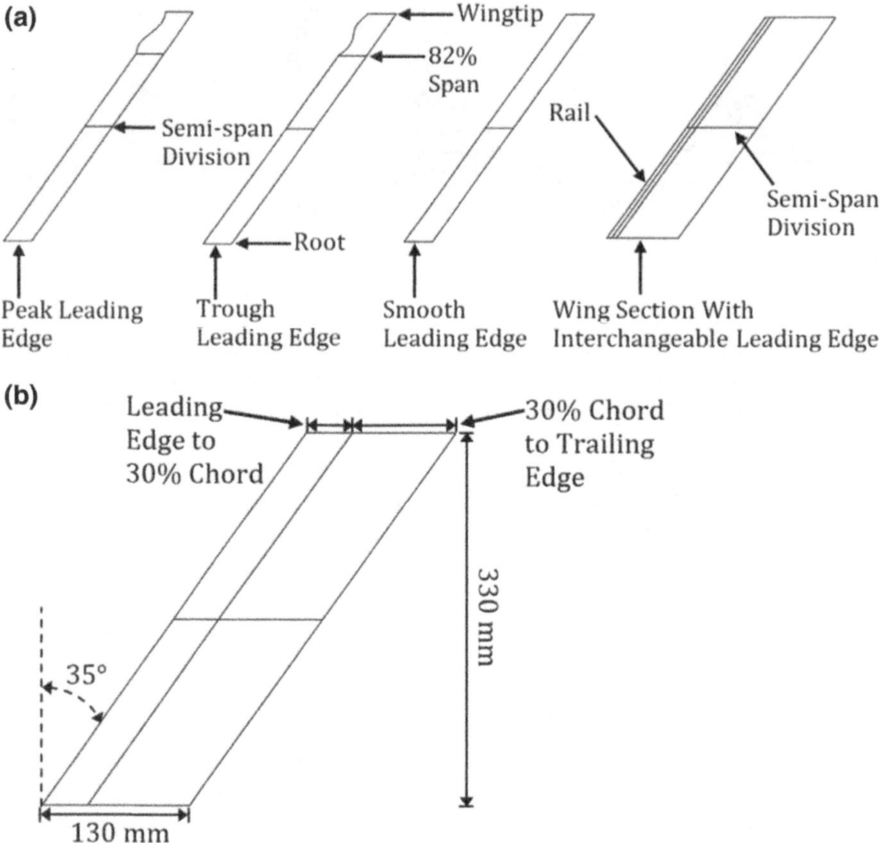

Fig. 7 Model used to investigate the effect of a single tubercle at the tip of a swept wing (Bolzon et al. 2017). **a** Wing section and three LE sections, **b** assembled smooth wing configuration. The term "Semi-Span Division" refers to the joins between parts of the wing assembly

tubercle producing a trough at the wing tip. For both tubercle cases A/c was 0.08 and λ/c of 0.46. Flow visualization and wake velocity surveys were conducted. Overall, for pre-stall AoAs up to 15°, the lift, drag and *lift-to-drag* ratio were within 3% of the values obtained for the smooth LE case. It was found that while the single wing-tip tubercle designs caused changes in the profile drag at the tip, induced drag effects were far smaller, and the wing tip vortex strength was typically changed by just 2.2%. Overall, the study concluded that neither design incorporating a single tubercle at the wingtip produced a significant performance change at pre-stall AoAs.

3 Alternative Geometries

Following the work by Fish and Battle (1995) on wings with tubercled LEs, Zverkov et al. (2008) modified the entire span of a foil with grooves and humps to impart an undulating appearance to it, as shown in Fig. 8. Flow visualization revealed that the boundary layer of the rippled foil exhibited different characteristics along the groove and hump spanwise locations. At a zero AoA, laminar separation bubbles formed along the grooves (troughs) leading to transition and subsequently turbulent flow re-attachment further downstream. In contrast, the flow remained attached along the humps (peaks). Zverkov et al. (2008) speculated that the modified foil would perform better than an unmodified one due to the resistance of its boundary layer to turbulent flow separation.

Arai et al. (2010) introduced a LE modification whereby half of a sinusoidal curve containing only the tubercle peak was incorporated into the design of wings, as shown in Fig. 9. Operating in a water tunnel at a Reynolds number of 138,000, the modified

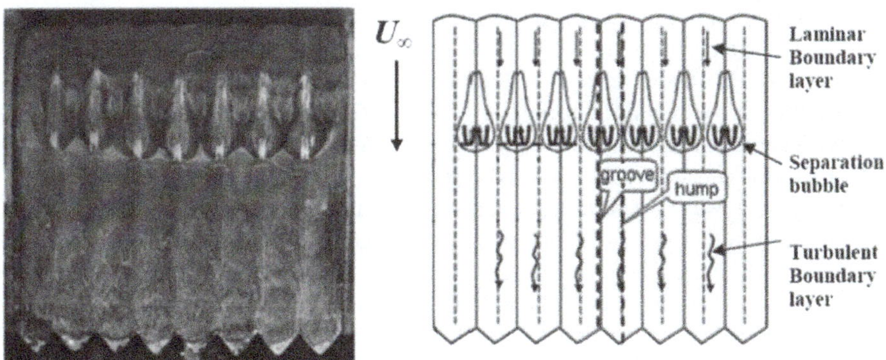

Fig. 8 A rippled foil producing three-dimensional laminar separation bubbles as demonstrated by Zverkov et al. (2008)

Fig. 9 Leading-edge tubercles designed based on half of a sinusoidal curve producing streamwise vortices as demonstrated by Arai et al. (2010)

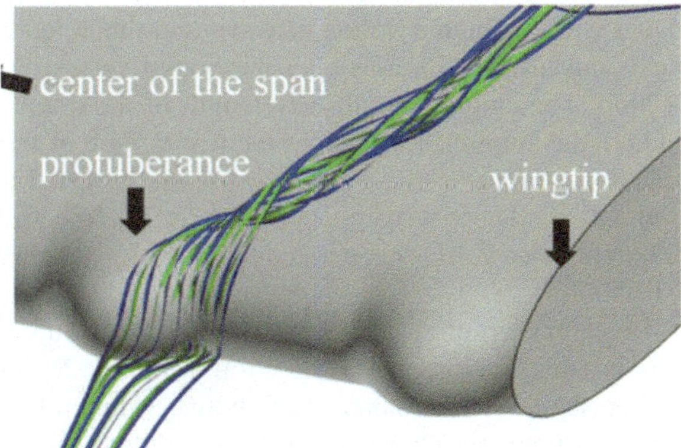

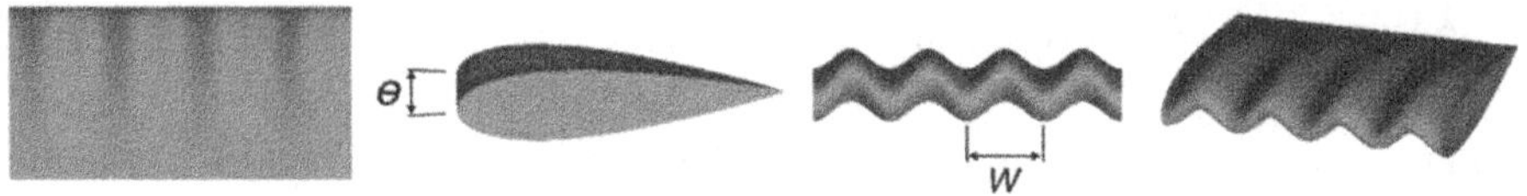

Fig. 10 The configuration of a wavy foil as illustrated by Rostamzadeh et al. (2013)

wings showed softer stall characteristics as, unlike the unmodified NACA 0018, the loss of lift was not abrupt around the stall angle. Furthermore, the modified wings produced streamwise vortices that prevented LE flow separation.

A novel alternative to LE tubercles, proposed by Kelso (2015) and investigated by Hansen (2012), is described in Fig. 10. Referred to as the wavy foil, the wing's LE is deformed such that the geometric attack angle varies sinusoidally along the span while the chord length remains unchanged. The concept design of the wavy foil was inspired by the recognition that tubercled and wavy LEs would induce a cyclic variation in the circulation associated with the wing's bound vortex along its span. Predicated on this notion, it was expected that tubercled and wavy lifting bodies would display similar aerodynamic loading behavior.

To assess the performance of wavy wings, Hansen (2012) conducted a series of wind tunnel tests on a number of airfoils with varying peak-to-peak amplitudes and wavelengths in the transitional regime ($Re = 120,000$). The baseline foil adopted for the design of the wavy foils was the NACA 0021, which closely resembles that of the humpback whale. It was found that, similar to tubercled foils, wavy foils achieved lower maximum lift coefficients in the pre-stall condition, but generated more lift in the post-stall zone when compared to the unmodified foil. Another similarity between the behavior of tubercled and wavy foils was ameliorated stall characteristics whereby, in contrast with their unmodified counterpart, the modified foils did not undergo a sudden loss of lift around the stall angle. Hansen (2012) reported that among the airfoils with the same angular amplitude, θ, those with smaller wavelengths, w, performed more favourably, whereas wavy foils with larger angular amplitudes outperformed others when the wavelength was kept unchanged.

Following the study by Hansen (2012), Rostamzadeh et al. (2013) adopted Prandtl's lifting-line model to analyse the spanwise distribution of circulation associated with the bound vortex of wavy and tubercled finite wings. It was found that wavy and tubercled wings alter the bound-vortex circulation in a spanwise-cyclic fashion. The amplitude and wavelength of the induced cyclic modulations depend on both the amplitude as well as the wavelength of the LE configuration. The cyclic variation of spanwise circulation was also reported in the numerical study by Cai et al. (2015). These researchers concluded that the upwash and downwash effects of the streamwise vortices generated by tubercles accounted for the cyclic spanwise lift distribution.

Rostamzadeh et al. (2013) carried out wind tunnel tests to analyse the surface pressure distribution over a wavy foil at $Re = 120,000$. This investigation revealed further shared aerodynamic traits between tubercled and wavy foils described as follows. Firstly, the chordwise pressure gradient was more adverse along the troughs compared with behind the peaks. Secondly, the suction peak (pressure minimum) within each trough was found to be higher in magnitude than that of each peak. Finally, the presence of counter-rotating streamwise vortices in the vicinity of the wavy wing was predicted by flow modelling.

Gawad (2012) incorporated spherical tubercles into the design of the LE of a NACA 0012 airfoil. In this study, flow modelling using numerical methods was utilized to assess the performance of the modified airfoil against one without tubercles at a range of Reynolds numbers (65,000–1,000,000). The investigation led to the conclusion that airfoils with spherical tubercles produced softer stall and generated more lift at high AoAs than the unmodified foil. Nonetheless, the unmodified foil produced more lift than the tubercled foils in the pre-stall zone. The behavior of foils with spherical tubercles is similar to that of foils with undulating LEs as reported in the literature (Hansen et al. 2011; Custodio 2007; Custodio et al. 2015).

Hasheminejad et al. (2017) modified the LEs of flat plates by three different configurations that incorporated periodic patterns described as triangular, sinusoidal and notched. The wavelength associated with the repeating patterns was set to 15 mm and the amplitude was maintained at 7.5 mm. Wind tunnel tests were conducted at a free-steam flow speed of 3 m/s. The development of streamwise vorticity over the plates was investigated using hot-wire anemometry. The results demonstrated that pairs of counter-rotating streamwise vortices formed downstream of the LE. The sense of rotation of the vortices was such that downwash and upwash were present behind the peak and the trough spanwise locations, respectively. This finding was in agreement with previous studies that focused on the vortex formation behind tubercled LEs (Custodio 2007; Hansen et al. 2016; Rostamzadeh et al. 2014; Weber et al. 2011). Whereas the vortex structure associated with the triangular and sinusoidal LEs were very similar, the notched LE produced an additional vortex structure behind the peaks that increases the instability of the boundary layer.

Mangeol et al. (2017) studied the impact of a serrated LE on the aerodynamic behavior of flat plates operating at Mach numbers from 0.2 to 0.64 at a Reynolds number of $12,000 \pm 1000$. Using the lift-to-drag ratio as a performance metric, the flat plate with larger-wavelength serrations marginally outperformed the baseline plate at around the stall angle, whereas the plate with smaller-wavelength serrations exhibited inferior loading characteristics. The Mach number was demonstrated to influence lift generation past the stall angle as the serrated plates produced more lift compared to the unmodified plate when the Mach number was increased. Nonetheless, in the pre-stall region, the Mach number did not alter the lift-curve slope. The serrated LEs gave rise to three main types of flow configurations with distinctive pressure distribution patterns that were periodic. These flow structures were also reported by Sakai et al. (2015). Overall, the serrated LE produced low pressure regions behind the troughs where the chordwise pressure gradients were steeper compared to those of the peaks, similar to the observations of Rostamzadeh et al. (2013) for the wavy wing configuration.

4 Conclusion

The majority of studies on full-span airfoils with various tubercle configurations have concluded that the best pre-stall aerodynamic performance can be achieved with small amplitude and small wavelength tubercles. In general, pre-stall airfoil performance was found to be inferior to that of the unmodified foils, although there were some exceptions (de Paula et al. 2017; Chong et al. 2015). The smallest amplitude-to-chord ratio that has been investigated so far is $A/c = 0.025$, and therefore there may be a limit to the performance benefits obtained by reducing the amplitude. The optimal choice of wavelength is dependent on the tubercle amplitude (Hansen et al. 2011) and the ideal ratio between amplitude and wavelength (A/λ) appears to be between 0.25 and 0.3, based on the available literature (Hansen et al. 2011; de Paula et al. 2017; Chong et al. 2015). Post stall, similar performance benefits were observed for the small-amplitude and small-wavelength configurations (Hansen et al. 2011; de Paula et al. 2017; Chong et al. 2015). The post-stall lift was found to be significantly higher for airfoils with tubercles compared to unmodified airfoils. Some results suggested that using larger amplitude tubercles improves the post-stall performance (Johari et al. 2007; Favier et al. 2012) and therefore further investigation is warranted. The limited body of data suggest that noise reduction is greatest for tubercles with small wavelength and large amplitude.

By contrast to the above, there is currently no consensus as to an optimum tubercle configuration for finite-span wings, although the trends of experimental studies suggest that smaller amplitude and smaller wavelength tubercles generally perform best in terms of lift, drag and efficiency (L/D). Two studies based on lifting-line theory indicate that there is little or no dependence on wavelength, although there was disagreement about the effects of amplitude. An artificial neural network study generated a definitive geometry but did not indicate any trends. A CFD study on a propeller suggested that larger tubercle amplitude and smaller wavelength led to a higher efficiency. Finally, one of the lifting-line studies demonstrated that a single tubercle trough located on the outboard end of the wing provided the best lift-to-induced-drag ratio. Furthermore, there is no clear trend in the effects of tubercle alignment. The strongest consensus relates to the optimum distribution of tubercles along the LEs of finite-span foils, where the evidence suggests that the tubercles are most effective when located adjacent to the wing tip, with a smooth inboard LE. Interestingly, the logical extension of this case—a single tubercle at the wing tip—has not shown any significant benefits from the configurations investigated thus far.

Finally, a number of different alternative geometries were identified, such as serrated and wavy wings, rippled wing surfaces and alternative versions of tubercles such as dome-shaped protrusions. All of these cases show significant similarities in their effects on the flow pattern next to the wing surface through to the lift and drag characteristics. This may suggest that these modifications, all of which produce a spanwise modulation in the wing's cross-sectional shape, constitute a family of geometries that produce a similar overall effect on wing performance.

References

Arai H, Doi Y, Nakashima T, Mutsuda H (2010) A study on stall delay by various wavy leading edges. J Aero Aqua Bio-mech 1(1):18–23

Biedermann TM, Chong TP, Kameier F, Paschereit CO (2017) Statistical–empirical modeling of airfoil noise subjected to leading-edge serrations. AIAA J 3128–3142

Bolzon MD (2016) The effects of tubercles on swept wing performance at pre-stall angles of attack. Ph.D. thesis, Mechanical Engineering, University of Adelaide, Adelaide, SA, Australia

Bolzon MD, Kelso RM, Arjomandi M (2016) Parametric study of the effects of a tubercle's geometry on wing performance through the use of the lifting-line theory. In: 54th AIAA Aerospace Sciences Meeting, p. 0295

Bolzon MD, Kelso RM, Arjomandi M (2017) Performance effects of a single tubercle terminating at a swept wing's tip. Exp Thermal Fluid Sci 85:52–68

Butt FR, Talha T (2018) A parametric study of the effect of the leading-edge tubercles geometry on the performance of aeronautic propeller using computational fluid dynamics (CFD). In: Proceedings of the World Congress on engineering

Cai C, Zuo Z, Liu S, Wu Y (2015) Numerical investigations of hydrodynamic performance of hydrofoils with leading-edge protuberances. Adv Mech Eng 7(7), 1687814015592088.Z

Chaitanya P, Joseph P, Narayanan S, Vanderwel C, Turner J, Kim JW, Ganapathisubramani B (2017) Performance and mechanism of sinusoidal leading edge serrations for the reduction of turbulence–aerofoil interaction noise. J Fluid Mech 818:435–464

Chong TP, Vathylakis A, McEwen A, Kemsley F, Muhammad C, Siddiqi S (2015) Aeroacoustic and aerodynamic performances of an aerofoil subjected to sinusoidal leading edges. In: 21st AIAA/CEAS aeroacoustics conference, p 2200

Custodio D (2007) The effect of humpback whale-like leading edge protuberances on hydrofoil performance. MS thesis, Worcester Polytechnic Institute

Custodio D, Henoch CW, Johari H (2015) Aerodynamic characteristics of finite span wings with leading-edge protuberances. AIAA J 53(7):1878–1893

de Paula AA, Meneghini J, Kleine VG, Girardi RD (2017) The wavy leading edge performance for a very thick airfoil. In: 55th AIAA Aerospace Sciences Meeting, p 0492

Favier J, Pinelli A, Piomelli U (2012) Control of the separated flow around an airfoil using a wavy leading edge inspired by humpback whale flippers. CR Mec 340(1–2):107–114

Fish FE, Battle JM (1995) Hydrodynamic design of the humpback whale flipper. J Morphol 225(1):51–60

Gawad AFA (2012) Numerical simulation of the effect of leading-edge tubercles on the flow characteristics around an airfoil. In: Proceedings of the ASME 2012 international mechanical engineering congress and exposition, pp 9–15

Gruber T, Murray MM, Fredriksson DW (2011) Effect of humpback whale inspired tubercles on marine tidal turbine blades. In: ASME 2011 International Mechanical Engineering Congress and Exposition, pp 851–857. American Society of Mechanical Engineers

Hansen KL (2012) Effect of leading edge tubercles on airfoil performance. Ph.D. thesis, Mechanical Engineering, University of Adelaide, Adelaide, SA, Australia

Hansen KL, Kelso RM, Dally BB (2010) An investigation of three-dimensional effects on the performance of tubercles at low Reynolds numbers. In: 17th Australasian fluid mechanics conference, Auckland, New Zealand, pp 5–9

Hansen KL, Kelso RM, Dally BB (2011) Performance variations of leading-edge tubercles for distinct airfoil profiles. AIAA J 49(1):185–194

Hansen KL, Rostamzadeh N, Kelso RM, Dally BB (2016) Evolution of the streamwise vortices generated between leading edge tubercles. J Fluid Mech 788:730–766

Hasheminejad SM, Mitsudharmadi H, Winoto SH, Low HT, Lua KB (2017) Development of streamwise counter-rotating vortices in flat plate boundary layer pre-set by leading edge patterns. Exp Thermal Fluid Sci 86:168–179

Johari H, Henoch CW, Custodio D, Levshin A (2007) Effects of leading-edge protuberances on airfoil performance. AIAA J 45(11):2634–2642

Kelso RM (2015) Improved wing configuration: Application Number WO/2014/026246, WIPO Patent

Mangeol E, Ishiwaki D, Wallisky N, Asai K, Nonomura T (2017) Compressibility effects on flat-plates with serrated leading-edges at a low Reynolds number. Exp Fluids 58(11):159

Rostamzadeh N, Kelso RM, Dally BB, Hansen KL (2013) The effect of undulating leading-edge modifications on NACA 0021 airfoil characteristics. Phys Fluids 25(11):117101

Rostamzadeh N, Hansen KL, Kelso RM, Dally BB (2014) The formation mechanism and impact of streamwise vortices on NACA 0021 airfoil's performance with undulating leading edge modification. Phys Fluids 26(10):107101

Sakai M, Sunada Y, Rinoie K (2015) Three-dimensional separated flow on a flat plate with leading-edge serrations. In: 53rd AIAA Aerospace Sciences Meeting, pp 0047

Shi W, Atlar M, Norman R, Aktas B, Turkmen S (2016a) Numerical optimization and experimental validation for a tidal turbine blade with leading-edge tubercles. Renew Energy 96:42–55

Shi W, Rosli R, Atlar M, Norman R, Wang D, Yang W (2016b) Hydrodynamic performance evaluation of a tidal turbine with leading-edge tubercles. Ocean Eng 117:246–253

Shi W, Atlar M, Norman R (2017) Detailed flow measurement of the field around tidal turbines with and without biomimetic leading-edge tubercles. Renew Energy 111:688–707

Taheri A (2018) A meta-model for tubercle design of wing planforms inspired by humpback whale flippers. World Academy of Science, Eng Technol Int J Aerosp Mech Eng 12(3)

Van Nierop EA, Alben S, Brenner MP (2008) How bumps on whale flippers delay stall: an aerodynamic model. Phys Rev Lett 100(5):054502

Weber PW, Howle LE, Murray MM, Miklosovic DS (2011) Computational evaluation of the performance of lifting surfaces with leading-edge protuberances. J Aircraft 48(2):591–600

Wei Z, New TH, Cui YD (2017) Aerodynamic performance and surface flow structures of leading-edge tubercled tapered swept-back wings. AIAA J 423–431

Yoon HS, Hung PA, Jung JH, Kim MC (2011) Effect of the wavy leading edge on hydrodynamic characteristics for flow around low aspect ratio wing. Comput Fluids 49(1):276–289

Zverkov I, Zanin B, Kozlov V (2008) Disturbances growth in boundary layers on classical and wavy surface wings. AIAA J 46(12):3149–3158

Flow Control by Hydrofoils with Leading-Edge Tubercles

T. H. New, Zhaoyu Wei, Y. D. Cui, I. Ibrahim and W. H. Ho

1 Introduction

Leveraging upon designs honed over millenniums by Mother Nature for engineering applications comes naturally to engineers who are interested in improving their performances as much as possible. In this chapter, the area of particular interest will be in marine engineering and applications that directly relate towards more efficient and better marine vessel manoeuvrability. Greater manoeuvre capabilities will bring about higher agility and better station-keeping abilities for these marine vessels. Furthermore, control surfaces tend to bring about parasitic drag, which should be kept

T. H. New (✉)
School of Mechanical and Aerospace Engineering, Nanyang Technological University, Singapore 639798, Singapore
e-mail: dthnew@ntu.edu.sg

Z. Wei
School of Oceanography, Shanghai Jiao Tong University, Shanghai 200240, China

Y. D. Cui
Temasek Laboratories, National University of Singapore, Singapore 117411, Singapore

I. Ibrahim
School of Engineering, University of Glasgow, Singapore 118227, Singapore

W. H. Ho
Department of Mechanical and Industrial Engineering, University of South Africa, Pretoria, South Africa

Present Address:
I. Ibrahim
Maritime Advisory, DNV GL Singapore Pte Ltd, Singapore 118227, Singapore

Present Address:
W. H. Ho
School of Mechanical, Aeronautical and Industrial Engineering, University of the Witwatersrand, Johannesburg, South Africa

at minimal levels for better fuel efficiency. Hence, there are significant motivations behind the notion that control surfaces could be designed based on what have already existed in nature, so as to fully exploit the potential benefits. As the use of surface vessels, unmanned underwater vehicles (UUV), autonomous underwater vehicles (AUV) and underwater gliders continue to proliferate; achieving this is especially pertinent towards better controllability and prolonged submerged times.

One recent idea that inspires better marine vessel control surface designs arises from an interesting observation that humpback whales possess relatively regular bumps or tubercles along the leading-edges (LE) of their pectoral flippers. With these whales executing agile turns and sharp rolls during preying, their surprisingly high manoeuvrability was attributed to these unique LE tubercles. Furthermore, dissected whale flippers show a relatively thick cross-section which ought to have given rise to significant flow separations. In particular, an early study on dissected whale flippers by Fish and Battle (1995) revealed that the cross-sections closely resemble lifting surfaces, with the LE tubercles postulated to act as effective flow control and stall-delaying devices. Hence, the notion that the LE tubercles could enhance manoeuvrability through mitigation of flow separation phenomenon and adverse pressure gradients, rapidly took hold among the research community (Fish and Lauder 2006; Fish et al. 2008). This prompted a plethora of experimental and numerical endeavours that subsequently saw a significant increase in our collective understanding of the favourable flow effects conferred by LE tubercles when they are imposed upon lifting and control surfaces such as aerofoils, hydrofoils, rudders and propellers, among others.

Systematic experimental investigations (Miklosovic et al. 2004, 2007; Murray et al. 2005; Johari et al. 2007; Custodio 2007; Goruney and Rockwell 2009; Hansen et al. 2011, 2016; Guerreiro and Sousa 2012; Zhang et al. 2013, 2014; Rostamzadeh et al. 2014; Chen and Wang 2014; Custodio et al. 2015; Wei et al. 2015b, 2016b; Bolzon et al. 2016; Chaitanya et al. 2017, just to name a few] show that LE tubercled wings are indeed capable of achieving better performance through enhanced flow control capabilities. It should be highlighted that, to ease systematic designs and result interpretations, most studies were carried out based on rectangular finite-wing configurations with regular sinusoidal tubercles. Interestingly, this also means that such configured tubercled wings share some similarities with wavy cylinders, with the exception that the waviness is confined only to the LE for the former. Miklosovic et al. (2004) observed up to 6% increments to the maximum lift and 40% improvements to the stall angle for a semi-span LE tubercled wing. Follow-on work by Murray et al. (2005) further demonstrated that LE tubercled whale flipper test models produced up to 9% more lift than when no LE tubercles were used. On the other hand, Johari et al. (2007) studied the effects of tubercle wave-length and wave-amplitude on a full-span wing and observed that the maximum lift coefficient decreases instead. Nonetheless, more gradual wing-stalls and up to 50% higher post-stall lift levels were achieved through LE tubercled wings during their study. Interestingly, flow visualization results from Johari et al. (2007) and Custodio (2007) show that LE tubercles function like small delta-wings, where they produce persistent streamwise-aligned counter-rotating vortex-pairs (CVPs). The presence of these streamwise CVPs is believed to re-energize the boundary layer along the suction surface and mitigate the occurrences of flow separation phenomenon, as is typically the case at progressively upstream locations along a relatively thick lifting surface (such as a whale flipper). This was soon validated by Stanway (2008) when particle-image velocimetry (PIV) measurements revealed that formations of streamwise CVPs are associated with delays in flow separations. It was also at around this time that Van Nierop et al. (2008) attempted to explain why LE tubercles are able to produce favourable flow effects by invoking aerodynamics-based Lifting-Line Theory.

More recent implementations of LE tubercles upon hydrofoils with a view to use them for marine engineering applications included those carried out by Weber et al. (2010), Fish et al. (2011a, b), Johari (2015), Ibrahim and New (2015), and Shi et al. (2016), just to name a few. The potential is significant, considering the various different types of flow control surfaces available in both surface and underwater marine vessels. Take for instance, surface vessels will typically have rudders aft of the propellers, while underwater vessels tend to have additional control surfaces such as bow-planes, sail-planes or tail-planes. It should also be highlighted that the tail-planes or rudders for underwater vessels are generally located upstream of the propellers, instead of downstream in almost all surface vessels. As such, the initial and boundary conditions governing these flow control surfaces can be very different and highly dependent upon the hull design, presence of flow separation associated with upstream structures, any swirl imparted by the propellers and possibility of cavitation, among other factors.

In the work reported by Weber et al. (2010), cavitation behaviour was reported to have been modified by the imposition of LE tubercles. In this case, cavitation initiates earlier and the onset of cavitation are sensitive towards the LE tubercles configuration. Additionally, lift and drag performances deteriorate between angles-of-attack (AoA) of $\alpha = 15°$ to $22°$, though lift performance recovers beyond $\alpha = 22°$. More interestingly, the study revealed that there may be a critical Reynolds number above which a selected LE tubercle geometry will not produce significant favourable effects. This strongly suggests that LE tubercle geometries have to be optimized for the end application in mind. Fish et al. (2011a, b) further emphasized the immense potential of leveraging upon LE tubercles as cost-effective and robust marine engineering passive flow control devices, through stall delay by modifying the state of boundary layers, reducing spanwise flows and tip vortex strengths. As cavitation phenomenon plays an important role in the efficacy of marine flow control surfaces, it does not come as a surprise that Johari (2015) conducted a flow visualization study to look at how LE tubercles affect the onset and distributions of cavitation bubbles along the hydrofoil suction surfaces. The investigation showed that cavitation tends to occur immediately downstream of the trough locations only, though the inception of cavitation occurs at a lower velocity than a hydrofoil without any LE tubercles. This signifies that, while cavitation behaviour may be better organized by LE tubercles, there may be a price to pay in terms of earlier cavitation formations, which would see the hydrofoil working at sub-optimal flow efficiency and acoustics levels.

Further extensive efforts to elucidate the flow behaviour of hydrofoils with LE tubercles were undertaken by the authors (New et al. 2014, 2015, 2016; Wei et al. 2015a, b, 2016a, b, 2017a, b, 2018) recently. In particular, the focus of these studies was on the use of tubercled hydrofoils at relatively low Reynolds numbers, such that they better represent scenarios at which smaller UUVs tend to operate. With swarming becoming more prevalent in marine engineering due to high redundancy levels, robustness and increasingly ease of access to key swarming technologies, the use of swarms of multiple small-sized UUVs to monitor oceans and marine life is expected to grow. However, the low Reynolds numbers also pose unique issues in terms of comparatively thicker boundary layers along flow control surfaces, greater tendency for flow separations, smaller lift-to-drag ratios in relatively stronger and unsteady ocean currents, among others. Earlier studies tend to emphasize more on high Reynolds number scenarios that are representative of larger-scale marine engineering applications and hence, flow data and insights upon much smaller Reynolds number scenarios are far scarcer. As such, these investigations seek to understand and address some of these unique flow challenges. In water-tunnel studies where different LE tubercled hydrofoils at a low Reynolds number of Re = 14,000 were conducted, it was observed that just like their higher Reynolds number counterparts, these hydrofoils demonstrate the ability to mitigate regular flow separations as compared to non-tubercled hydrofoils. Instead of producing persistent flow separations that led to significant growth in the separation bubble along the suction surface, tubercled hydrofoils tend to produce less and milder flow separations at trough locations. More importantly, despite the low Reynolds number, streamwise CVPs are formed aft of the LE tubercles at up to $\alpha = 20°$. Nonetheless, their persistence along the suction surface reduces as the angle-of-attack increases, due to the progressively higher adverse pressure gradient. It was further observed that the extent to which flow separation was mitigated is sensitive towards the exact LE tubercle configuration. Tubercle wave-amplitude was found to confer more appreciable influences than its wave-length, where the former was observed to have negligible flow separation differences below a certain value. Interestingly, experimental results also showed that a larger wave-amplitude leads to a corresponding larger effective tubercle height-to-boundary-layer-thickness ratio, h_{eff}/δ, the latter of which deviates from values associated with typical vortex generators (i.e. 0.1–0.5) (Lin 2002). Hence, this suggests that for LE tubercles used in scenarios whereby the height-to-boundary-layer-thickness ratios exceed 0.5 significantly, their favourable effects could be further attributed to the modifications of the streamwise and spanwise pressure gradients, as well as any additional vortex structures produced by the LE tubercles.

Due to the prominent role played by streamwise CVPs in LE tubercled hydrofoils, their stability and robustness will be of significant interest, especially if LE tubercles were to be implemented upon high-lift hydrofoils that are relatively thicker or more cambered. This was investigated by New et al. (2016) where LE tubercled hydrofoils based on SD7032 profiles were tested. Experimental results demonstrated that the streamwise CVPs are considerably unsteady but tend to remain close to the suction surface at small AOAs and/or up to a moderate distance downstream of the LE tubercles. The use of larger AoAs or camber induces stronger adverse pressure gradients that will cause the streamwise CVPs to be displaced further away from the suction surface and meander along the spanwise direction—behaviour which could render the use of LE tubercles far less effective or even incur poor flow coherence that lead to worse performance than a hydrofoil without any LE tubercles. Supporting observations were also been made by Wei et al. (2016b, 2017a) when Proper Orthogonal

Decomposition (POD) technique was used on PIV measurements to look into how the instantaneous flow behaviour and energy distributions could be altered through the use of differently configured LE tubercles. An interesting observation lies in how the use of LE tubercles are able to reduce the flow energies associated with modes 1 and 2 large-scale wake vortex-shedding typically produced by a non-tubercled hydrofoil at $\alpha = 0°$. As the AOA increases however, such a behaviour gradually reduces. In particular, the LE tubercled hydrofoil actually possesses higher flow energy levels for mode 1 vortex-shedding than its non-tubercled counterpart at $\alpha = 20°$, demonstrating again that LE tubercles tend to work well up to a certain extent before the benefits are negated or even reversed. Additionally, critical points were identified in the PIV measurements (similar to earlier studies), where further analysis suggests that they are associated with recirculating flows aft of the troughs.

In this chapter, further insights from recent efforts on LE tubercled hydrofoils by the authors will be presented. In particular, two new developments will be reported—Firstly, numerical approaches have now been adopted to complement experimental work and secondly, the flow behaviour of such hydrofoils operating under dynamic motion is now being considered. Numerical simulations complement experimental measurements and enable far more detailed flow behaviour to be isolated and identified. On the other hand, LE tubercled hydrofoils undergoing dynamic motions resembling abrupt turning/veering manoeuvres have not been studied much and it is hoped that some of the results here will prompt further studies to tackle this challenge. The next section will cover the experimental/numerical setups and procedures implemented, followed by a section discussing upon the latest experimental insights on LE tubercled hydrofoils. Subsequently, numerical results for similarly configured hydrofoils under both static and dynamic configurations will be presented over the next two sections and compared with existing experimental results. Lastly, these latest findings will be summarized and conclusions on some possible future research directions will be proposed.

2 Experimental Approaches and Hydrofoil Designs

All experiments were conducted in a 450 mm (W) $\times$ 600 mm (H) $\times$ 1100 mm (L) test-section of a low-speed recirculating water tunnel at Nanyang Technological University, as shown in Fig. 1. The two sides and bottom wall of the water tunnel were fabricated from tempered glass and hence, provided good optical access for particle-image velocimetry measurements. Additionally, a 40 mm (W) $\times$ 250 mm (H) tempered glass window was available at the back of water tunnel for cross-stream measurements. Flow was conditioned by several honeycomb structures and fine-mesh screens before entering a 4:1 contraction section to provide a free-stream with low turbulence intensity (~1.1% at a working velocity of $U_o = 0.19$ m/s). The uncertainty in the free-stream Reynolds number was also approximately $\pm 1.2\%$. All experiments were conducted at a Reynolds number of $Re = U_o c/\nu = 14,000$, where c is the mean hydrofoil chord and ν is water kinematic viscosity under test conditions.

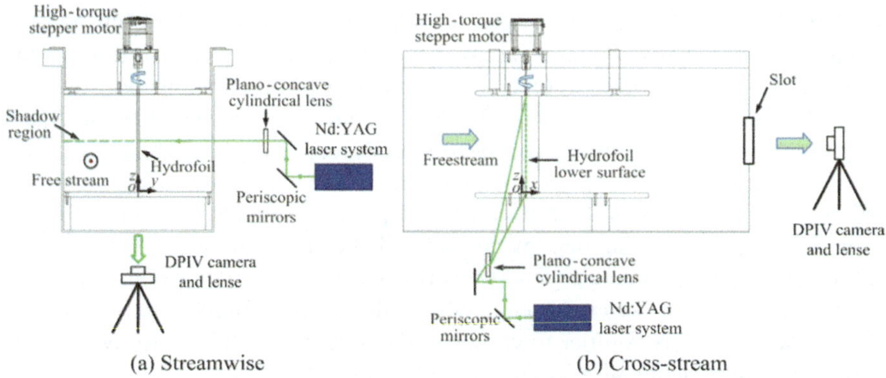

Fig. 1 Schematics of the experimental setup for **a** streamwise and **b** cross-stream DPIV measurement experiments. Adapted from Wei et al. (2015b)

As illustrated in Fig. 1, two flat transparent acrylic plates were installed beneath the water surface, such that the test hydrofoils could be installed between these two plates. The upper plate was covered by thin black adhesive papers to minimize laser reflections, as well as to provide a dark background during PIV experiments to better contrast against the illuminated particles. To adjust the AoA, a computer-controlled high-torque stepping motor was attached to the upper plate, with its shaft coupled to the test hydrofoil. Looking from the top, the AoA was positive in the clockwise direction.

The experiments were based on cambered SD7032 and FX63-137 profiles, where they are generally considered to be "high-lift" profiles and considerably less well-studied as compared to non-cambered NACA634-021 or NACA0020 profiles. In fact, much of the existing knowledge regarding tubercled hydrofoils came from the use of non-cambered profiles and the impact of LE tubercles on cambered hydrofoils is less explored. All test hydrofoils used for the experiments were fabricated from aluminium, with a consistent mean chord and span of $c = 75$ mm and $s = 300$ mm respectively. Two different tubercle geometries were considered for each profile, where a constant wave amplitude of $A = 9$ mm (0.12c) was used with two different wavelengths of $\lambda = 18.75$ mm (0.25c) and 37.5 mm (0.5c). This provided two distinctive amplitude-to-wavelength ratios of $A/\lambda = 0.48$ and 0.24 for comparison purposes. Details of the four different test hydrofoils are listed in Table 1. These wave amplitudes and wavelengths were selected as they fall within the range of

Table 1 Configuration details of the four different test hydrofoils

SD7032	FX63-137	Wave amplitude (mm)	Wavelength (mm)
Baseline		0	0
A9λ18.75		9	18.75
A9λ37.5		9	37.5

values associated with humpback whale flippers (Fish and Battle 1995; Johari et al. 2007).

The design methodology is relatively similar to those adopted by Lohry et al. (2012) and Wei et al. (2015b) for non-cambered LE tubercled hydrofoils, with the exception of certain differences to take in account the camber. In this case, a fifth-order polynomial curve was used to fit to the hydrofoil camber as

$$f(x)_{camber} = a_0 x^5 + a_1 x^4 + a_2 x^3 + a_3 x^2 + a_4 x + a_5 \tag{1}$$

This polynomial curve was extrapolated till the location, x/c = −0.12, where it was also the start-point of the final hydrofoil camber. To maintain the maximum thickness position and the hydrofoil section behind the maximum thickness point, the following nonlinear transformations in both the x and y directions were used:

$$x_1 = \begin{cases} x + 0.5 x_{LE}[1 + \cos(\pi x/e)] & 0 \le x < e \\ x & x \ge e \end{cases} \tag{2}$$

$$y_1 = \begin{cases} y + 0.5 y_{LE}[1 + \cos(\pi x/e)] & 0 \le x < e \\ y & x \ge e \end{cases} \tag{3}$$

where x and y are the coordinates along the baseline profile, x_1 and y_1 are the coordinates of the profile with LE tubercles after transformation, and e is the maximum thickness position. y_{LE} was determined using Eqs. (1)–(3) to satisfy the boundary conditions and continuities of two sections over the maximum thickness location. Figure 2 shows a comparison of the cross-sectional profiles located at the peak and trough locations against the baseline cambered profile.

To evaluate the hydrodynamic characteristics of tubercled hydrofoils, PIV measurements along the streamwise and cross-stream directions were conducted. Velocity and vorticity flow information derived from these measurements provided both qualitative and quantitative appreciation of the resulting flow behaviour. A 15 Hz, 200 mJ/pulse, Litron double-pulsed Nd:YAG laser was used as the illumination source, where beam-steering and sheet-forming optics were used to generate thin light sheets. Instantaneous particle images were then captured by a 1600 × 1200 pixels Dantec Dynamics FlowSense 2 M/E double-frame CCD camera with a Nikon f/1.4, 50 mm lens. Image-capturing was synchronized with the operations of the

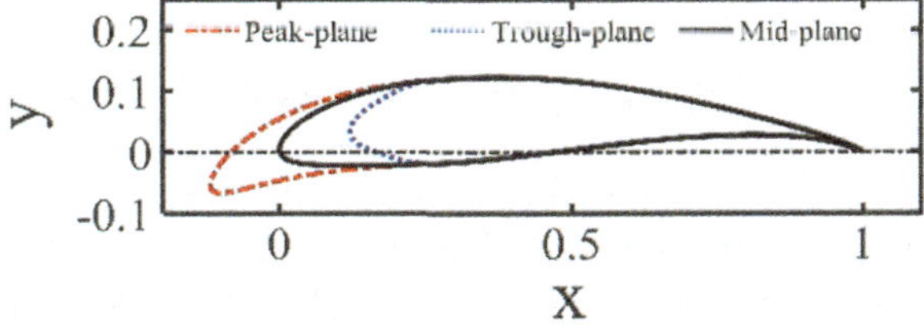

Fig. 2 Comparison of the normalized cross-sections located at the peak, mid and trough locations for FX63-137 LE tubercled hydrofoils

Nd:YAG laser through a Dantec Dynamics PIV Timer Box connected to a workstation equipped with appropriate timer and image-grabber boards. For streamwise measurements, they were taken along the Trough-, Mid- and Peak-planes (i.e. abbreviated as T-, M- and P-planes hereafter) of the test hydrofoils. 615 image-pairs were captured for a physical measurement window of approximately 120 mm $\times$ 90 mm at 15 Hz for each AoA to ensure satisfactory convergence in the mean flow field. On the other hand, cross-stream measurements utilized the same camera but with a Nikon f/2, 105 mm lens instead. 1000 image-pairs at 15 Hz were captured for a measurement window of approximately 130 mm $\times$ 100 mm to ensure satisfactory convergence as well.

For post-processing, a 2-step refinement multi-grid cross-correlation technique with initial and final interrogation window sizes of 128 $\times$ 128 pixels and 32 $\times$ 32 pixels were used respectively. An overlapping ratio of 50% in both vertical and horizontal directions was also used, which led to 99 $\times$ 74 vectors in each velocity map. Each velocity map was subjected to both global and local rejection criteria, where spurious vectors would be replaced by a 3-point by 3-point neighbourhood average. Resolution of the velocity maps was estimated to be approximately 1.25 mm/vector and vorticity maps were subsequently determined via a central-differencing scheme. Based on the measurement and post-processing procedures, the uncertainty levels of the measured velocity components were limited to within $\pm 1\%$ (Keane and Adrian 1992). Furthermore, taking the test conditions and experimental setup into account would yield a total experimental uncertainty of approximately 3.6% and 4% in the velocity and vorticity results (Moffat 1988).

For the numerical simulations, LE tubercled hydrofoils based on NACA634-021 A9λ37.5 profiles were investigated to complement the earlier studies conducted on a similarly configured LE tubercled hydrofoil by Wei et al. (2015b, 2016b) and Joy et al. (2016). Similar to the hydrofoils described in the previous section, the test hydrofoil used in the simulations has a mean chord and span of c = 75 mm and s = 300 mm respectively for the sake of consistency with the earlier experimental studies. The LE tubercles possessed a wave-amplitude and wave-length of A = 9 mm and λ = 37.5 mm respectively.

For the NACA634-021 LE tubercled hydrofoils used in the present simulations, a Cartesian coordinate system has its origin tagged to the lower end of the mean LE, where x is on the lower end face pointing towards the trailing-edge (TE), z is on the LE pointing upwards, and y is determined by the right-hand law. With the hydrofoil TE defined at x_{TE} = c, the tubercled LE with n number of sinusoidal waves can then be described as

$$x_{LE} = A \sin(2n\pi z/s), \quad 0 \leq z \leq s \tag{4}$$

To create the different 2D cross-sections that made up the tubercled LE, a nonlinear shearing transformation was implemented, such that the baseline hydrofoil LE radius, maximum thickness position, profile behind maximum thickness point and continuity of the cross-section profile over maximum thickness point were preserved. The imposed transformation functions are

$$x_1 = \begin{cases} x + 0.5x_{LE}[1 + \cos(\pi x/0.3c)] & 0 \le x < 0.3c \\ x & x \ge 0.3c \end{cases} \tag{5}$$

where x and x_1 are the abscissa of the baseline profile and modified cross-section respectively. The derivatives of Eq. (5) can then be expressed as

$$\frac{dx_1}{dx} = \begin{cases} 1 - 5x_{LE}\pi \, \sin(\pi x/0.3c)/3c & 0 \le x < 0.3c \\ 1 & x \ge 0.3c \end{cases} \tag{6}$$

Note that Eq. (6) maintains the LE radius, smoothing of the two sections over maximum thickness point and boundary conditions. Furthermore, the hydrofoil is symmetrical about the mid-span with both ends at the trough location of the sine wave. Two different flow scenarios were considered for the simulations—one where the hydrofoil is immersed in a free-stream statically at high AoAs, while another where it is oscillating in a free-stream at a moderate absolute frequency and amplitude.

For the static scenario, the geometry and dimensions (in mm) of the computational domain are shown in Fig. 3a. The inlet boundary was denoted as the curved surface of the simulation domain, while the outlet boundary was located at 24c downstream of the hydrofoil. A symmetry boundary condition was prescribed at both sides of the domain. Discretization of the computational domain was through the use of a structured, hexahedral grid that accommodated the complex three-dimensional geometry of the model. As shown in Fig. 3b, the mesh consisted of approximated 9,000,000 elements and produced an overall y^+ value of less than 1 along the hydrofoil surfaces, where the inflated zones were expected to produce better resolutions for flows close to the hydrofoil surfaces. The simulations were conducted using ANSYS® Fluent software, based on unsteady Reynolds Averaged Navier-Stokes (RANS) with Transitional Shear Stress Transport (SST) turbulence model implemented. Similar to Rostamzadeh et al. (2014), this turbulence model had been selected due to the expectation that transitional flow features such as laminar separation bubbles (LSB)

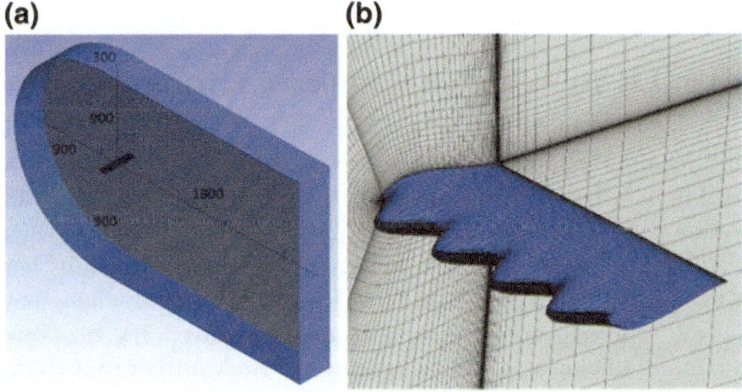

Fig. 3 a Schematics and dimensions, as well as **b** discretization of the computational domain (showing the mesh along the mid-span plane)

would be formed along the hydrofoil surfaces. A free-stream turbulence intensity of 1.1% and time-step defined at 0.001 s were prescribed for all computations. The simulations were carried out at Re = 14,000 at AoAs of $\alpha = 20°$, $30°$ and $40°$.

As for the oscillating tubercled hydrofoil scenario, the computational domain and mesh setup followed those used by Benaissa et al. (2017) and Flint et al. (2017), and were relatively similar to those adopted for the static scenario, except that a circular internal "cell-zone" was created around the hydrofoil to allow for its oscillation using sliding-mesh method. The overall mesh had approximately 10,000,000 elements with an overall y+ value of less than 10 in most parts of the domain, especially around the hydrofoil. Simulations of the desired oscillating motion can be achieved with either the sliding or dynamic mesh model within Fluent. However, sliding mesh (or mesh motion) was used here because it required less computational resources as there was no re-meshing taking place with the motion. A user-defined function (UDF) was used to rotate the circular "cell-zone" clockwise and anticlockwise periodically to simulate the hydrofoil oscillation motion. When the hydrofoil was increasing its AoA, it is termed as "pitch-up" and vice versa. The sliding mesh model required the boundary shared between the rotating "cell-zone" and the rest of the boundary to be set as matching interfaces. A point of caution to note when using the sliding mesh method is the formation of random "artificial flow features" at the sliding interface due to the mismatch in the mesh during certain parts of the motion. Occasionally, these flow features could propagate towards the hydrofoil and causes convergence problems. However, this can be circumvented by increasing the size of the internal "oscillating" cell-zone or by decreasing the mesh size at the matching interface. Note that either measure would inevitably increase the computational requirements. The Scaled Adaptive Simulation (SAS) turbulence model (Taylor et al. 2003) with PRESTO! Scheme was used in this numerical study. All other variables were calculated using second-order upwind discretization and the gradients evaluated using the cell-based least squares method. A double precision solver and second-order implicit transitional formulation was also used. Time-step used in this study was 0.004 s with a convergence criteria of 10^{-5}.

3 Effects of LE Tubercles on Cambered Hydrofoils

Experimental results for both SD7032 and FX63-137 baseline hydrofoils are first presented in Fig. 4, where mean streamwise vorticity fields and streamlines taken along their mid-spans at four AoAs of $\alpha = 0°$, $10°$, $15°$ and $20°$ are shown. Figure 4a shows that adverse pressure gradients due to the cambered hydrofoils lead to flow separations even at $\alpha = 0°$ at the present low Reynolds number, where they occur at approximately 2/3 chord location. As the AoA increases to $\alpha = 10°$, the flow separates from the maximum thickness point. Increasing the AoA further to $\alpha = 15°$ and $20°$ will lead to flow separations right from the LE that produce pronounced recirculating regions. Hence, stall angles for both hydrofoil designs are approximately $\alpha = 15°$. It is also interesting to note that recirculating regions for the FX63-137 hydrofoil tend

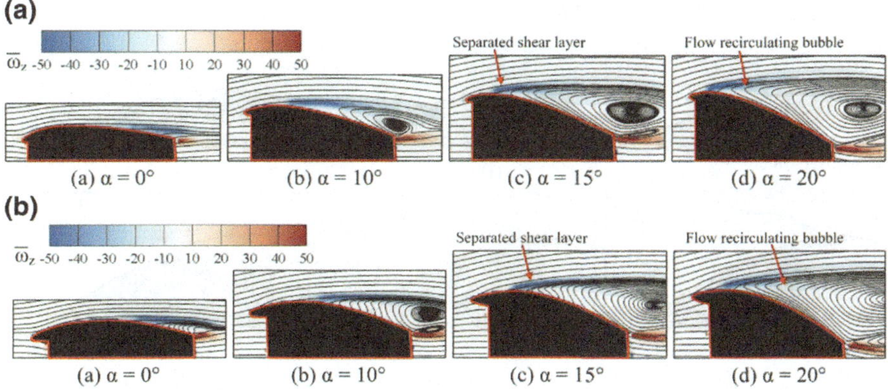

Fig. 4 Mean streamwise vorticity fields and streamlines for **a** SD7032 and **b** FX63-167 baseline hydrofoils along their mid-spans up to $\alpha = 20°$

to be more pronounced than those for the SD7032 hydrofoil throughout. It is clear from these results that such cambered hydrofoils will stall more readily at lower Reynolds number, which is highly undesirable from a practical standpoint. When LE tubercles are introduced however, flow separation behaviour begins to deviate significantly and demonstrates sensitivity towards the exact spanwise location. To illustrate the behaviour here, selected results at $\alpha = 10°$ and $20°$ (i.e. pre- and post-stall conditions) will now be presented and discussed.

Starting with SD7032 A9λ37.5 hydrofoil shown in Fig. 5, it can be discerned that the flow always separates from the LE along the T-plane in both pre- and post-stall regimes. However, the extent to which that happens is far greater at $\alpha = 20°$, with a node of flow detachment forming along the hydrofoil surface. As one moves towards

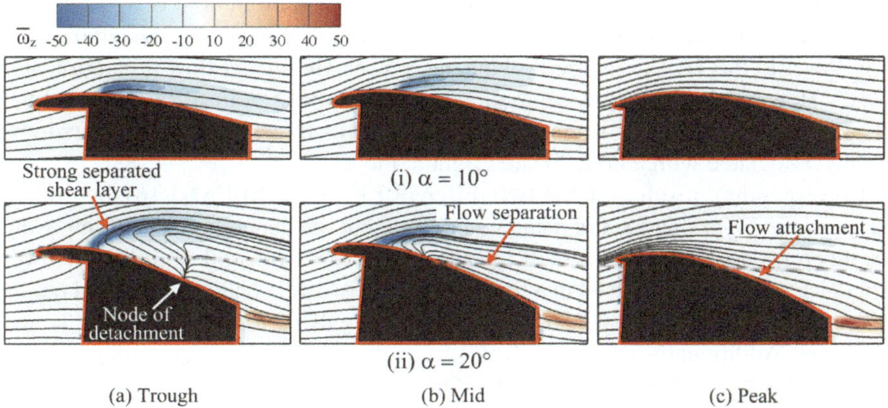

Fig. 5 Mean streamwise vorticity fields and streamlines for SD7032 A9λ37.5 hydrofoil along **a** T-, **b** M- and **c** P-planes at $\alpha = 10°$ and $20°$

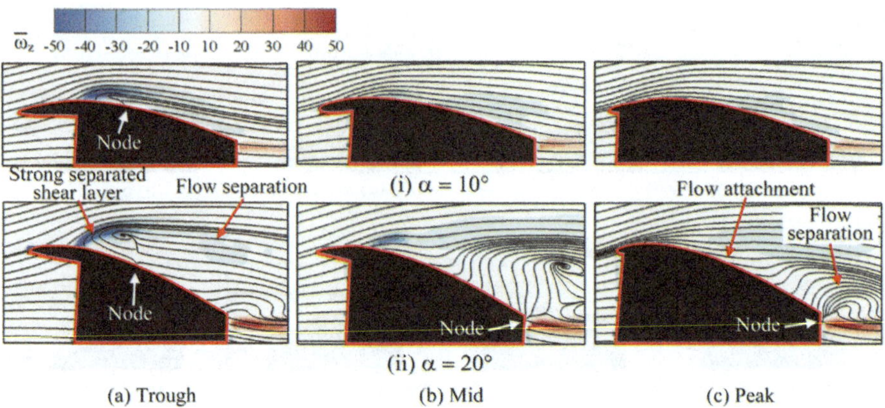

Fig. 6 Mean streamwise vorticity fields and streamlines for SD7032 A9λ18.75 hydrofoil along **a** T-, **b** M- and **c** P-planes at α = 10° and 20°

the M-plane, larger and smaller flow separations occur at α = 10° and 20° respectively. A flow detachment node is also present α = 20°, though it is further upstream than that along the T-plane. As described by Wei et al. (2016b), relatively higher and lower pressures are produced along the P- and T-planes respectively, which leads to flows diverting from the former to the latter. As a result, flows from this critical point create the recirculating flow as they move upstream. When the tubercle numbers are doubled however (i.e. halving the tubercle wave-length) as shown in Fig. 6, the most visible flow changes occur at post-stall conditions. Nevertheless, at pre-stall conditions, a node of detachment is produced even at α = 10° along the T-plane. At α = 20°, a saddle and a focus can also be observed, which indicate more complicated flow behaviour and structures when smaller tubercle wave-lengths are used. Moving to the M-plane, the separated shear layer become relatively weaker and no flow separation occurs at α = 10°. At α = 20° however, a flow separation bubble is formed from about half-chord location onwards. Critical points can be observed, though they are now closer to the TE. Along the P-plane, there is no significant flow separation at α = 10°. However, a small flow separation bubble can be seen to form just aft of the TE with a very weak separated shear layer. Lastly, the flow separation appears to be closely associated with a node located right at the TE.

Corresponding results for the two different LE tubercled FX63-167 hydrofoils are shown in Figs. 7 and 8 and it can be inferred that the general effects of imposing LE tubercles are quite similar. For instance, for a test hydrofoil with larger wave-length LE tubercles immersed in a free-stream at post-stall conditions, critical points indicative of a more three-dimensional flow field will be formed along the T- and M-planes. Additionally, there is a lack of significant flow separation along the P-plane at the same post-stall condition. On the other hand, using a test hydrofoil with smaller wave-length LE tubercles will produce flow separation bubbles that are delayed further along the hydrofoil surface along the T- and M-planes, to about half-chord location and beyond at post-stall conditions. Again, no strong flow separations

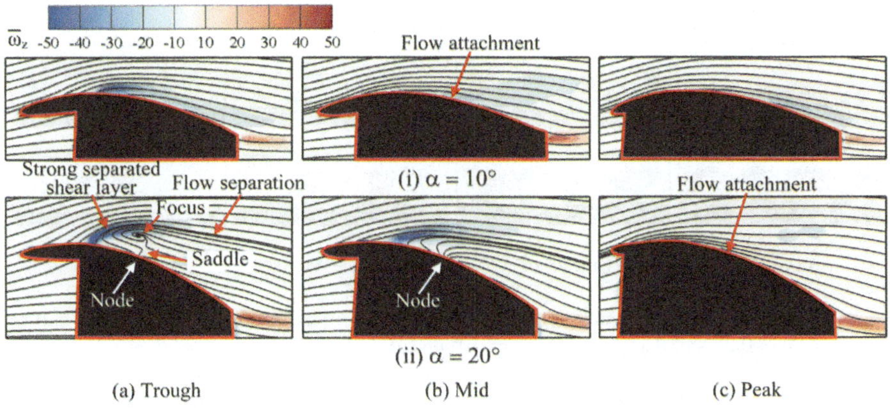

Fig. 7 Mean streamwise vorticity fields and streamlines for FX63-167 A9λ37.5 hydrofoil along **a** trough, **b** mid and **c** peak planes at $\alpha = 10°$ and $20°$

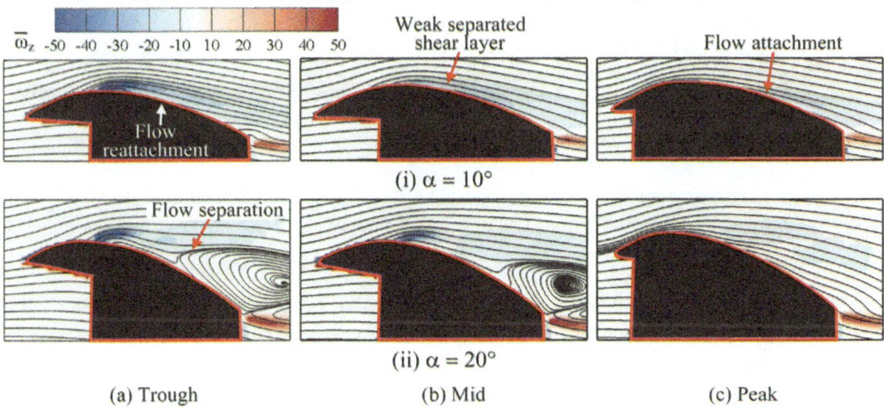

Fig. 8 Mean streamwise vorticity fields and streamlines for FX63-167 A9λ18.75 hydrofoil along **a** T-, **b** M- and **c** P-planes at $\alpha = 10°$ and $20°$

can be detected along P-plane at the same flow setting. Comparing to the results produced by the baseline hydrofoils, it is clear that the use of LE tubercles upon cambered hydrofoils continues to benefit mitigation of flow separation, even when comparatively more adverse pressure gradients exist along the hydrofoil suction surfaces.

Figures 9 and 10 show the mean cross-stream vorticity fields for the two tubercled SD7032 hydrofoils (i.e. A9λ37.5 and A9λ18.75) at $x/c = 0.12, 0.25, 0.38$ and 0.52 locations under $\alpha = 10°$ and $20°$ configurations. The physical width of the measurement window was 75 mm, which spanned over two and four wave-lengths for the A9λ37.5 and A9λ18.75 test hydrofoils respectively. Results indicate that, regardless of the AoA used, a pair of streamwise CVPs is produced by each LE tubercle in

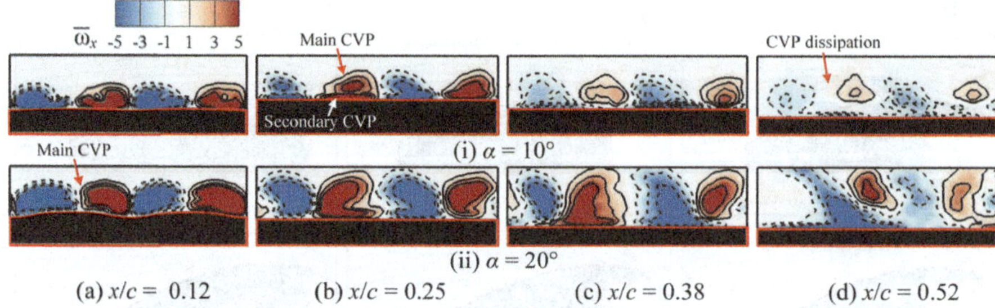

Fig. 9 Mean cross-stream vorticity fields for SD7032 A9λ37.5 hydrofoil along **a** $x/c = 0.12$, **b** $x/c = 0.25$, **c** $x/c = 0.38$ and **d** $x/c = 0.52$ at $\alpha = 10°$ and $20°$

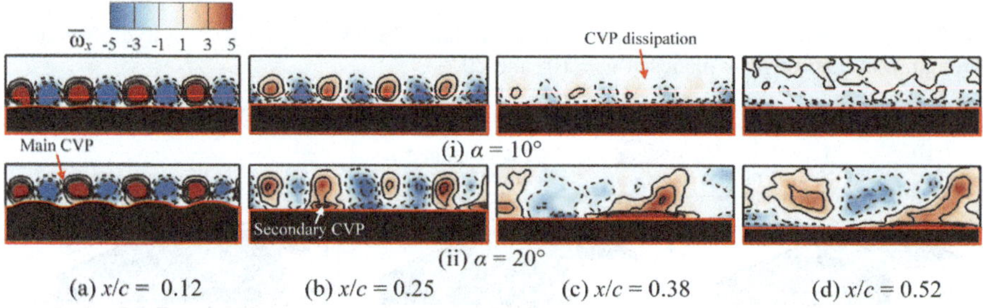

Fig. 10 Mean cross-stream vorticity fields for SD7032 A9λ18.75 hydrofoil along **a** $x/c = 0.12$, **b** $x/c = 0.25$, **c** $x/c = 0.38$ and **d** $x/c = 0.52$ at $\alpha = 10°$ and $20°$

both test hydrofoils. Interestingly, it can also be observed that the robustness of the streamwise CVPs is more obvious for A9λ37.5 test hydrofoil, where they persist in a more regular fashion up to x/c = 0.52. In contrast, those produced by A9λ18.75 test hydrofoil appears to have largely dissipated by x/c = 0.38 location. Furthermore, their vortex strengths also gradually reduce as they convect further along the hydrofoil suction surfaces. These observations agree well with New et al. (2016), where time-resolved PIV measurements led them to postulate that shorter wave-lengths lead to more intense interactions between neighbouring streamwise CVPs, which subsequently hasten their unsteadiness and viscous dissipation.

Corresponding results for tubercled FX63-137 test hydrofoils are presented in Figs. 11 and 12 and it is interesting to note that the streamwise CVPs produced by the FX63-137 A9λ37.5 hydrofoil are not as resilient as its SD7032 counterpart. For instance, it is clear that they become more incoherent from x/c = 0.38 location onwards, regardless at pre- or post-stall condition. On the other hand, however, those produced by the FX63-137 A9λ18.75 hydrofoil (i.e. with half the wave-length) maintained a higher level of coherence even at x/c = 0.52 location at pre-stall condition. At post-stall conditions, they become comparatively incoherent by x/c = 0.38 location, similar to the SD7032 A9λ18.75 hydrofoil.

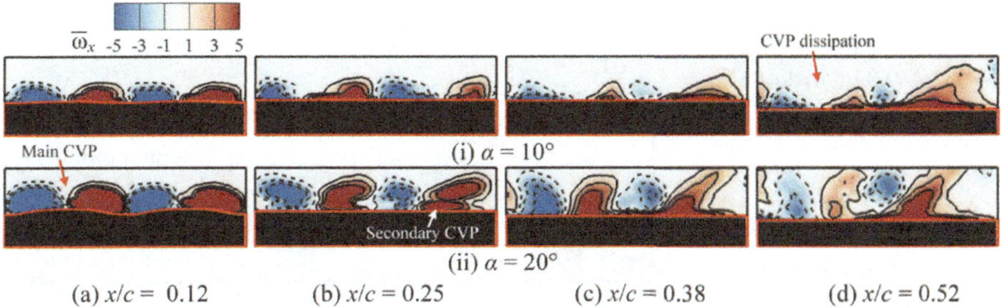

Fig. 11 Mean cross-stream vorticity fields for FX63-167 A9λ37.5 hydrofoil along **a** x/c = 0.12, **b** x/c = 0.25, **c** x/c = 0.38 and **d** x/c = 0.52 at α = 10° and 20°

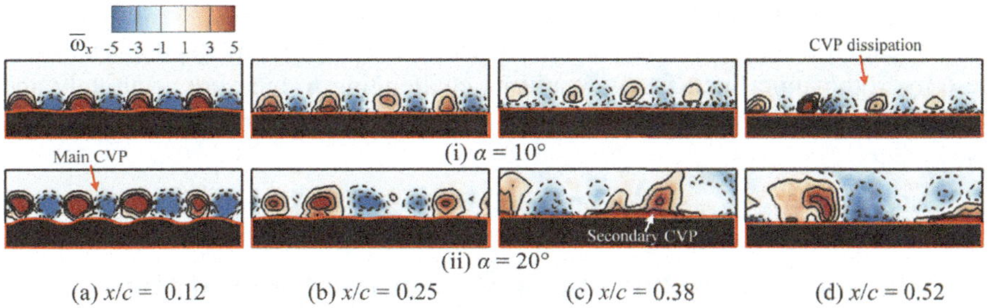

Fig. 12 Mean cross-stream vorticity fields for FX63-167 A9λ18.75 hydrofoil along **a** x/c = 0.12, **b** x/c = 0.25, **c** x/c = 0.38 and **d** x/c = 0.52 at α = 10° and 20°

As can be discerned from the preceding results, the formation of the streamwise CVPs is the key flow difference between the baseline and LE tubercled hydrofoils here, regardless of the exact profile used. Collating with the flow separation behaviour along different spanwise planes, it can be speculated that the streamwise CVPs are at least partially associated with the attached flows downstream of the LE tubercle peaks. Combined with earlier studies by Stanway (2008), Rostamzadeh et al. (2014) and Wei et al. (2015b, 2018), it is quite clear that LE tubercles function well in both cambered and non-cambered hydrofoils, as deduced through the skewed flow streaks by Rostamzadeh et al. (2014) and the tubercle-as-delta-wing analogy by Wei et al. (2015b) previously. Downwash and spanwise flows associated with both streamwise CVPs and critical points enhance fluid momentum transport from the peaks to the troughs of the LE tubercles, thus energizing the boundary layer and delaying the flow separation (Rostamzadeh et al. 2014; Skillen et al. 2015).

Before concluding this section, it is worthwhile to take a closer look at the highly irregular flow fields produced by unsteady streamwise CVPs and flow separations associated at high AoAs. Take for instance the case of SD7032 A9λ37.5 hydrofoil at α = 20° along x/c = 0.52 location presented in Fig. 13, it can be observed that the vortex structures formed downstream of the LE tubercle peaks and troughs are irregular. While mutual interactions between the streamwise CVPs are postulated to be one of

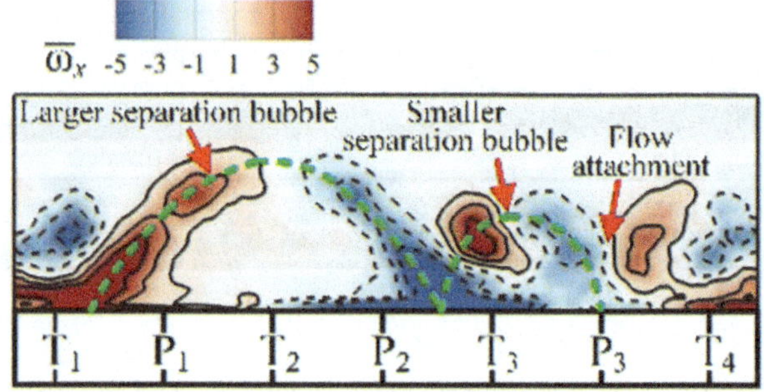

Fig. 13 Highly irregular flow field produced by SD7032 A9λ37.5 hydrofoil at $\alpha = 20°$ along x/c = 0.52 location

the causes of their unsteadiness, interactions between these streamwise CVPs and flow separations can further exacerbate the overall flow instability, possibly leading to significant fluctuating hydrodynamic forces. At the present low Reynolds number, these hydrodynamic fluctuations would be challenging to measure accurately and hence, they should be studied further under flow conditions that permit more accurate assessments.

4 Formation and Behaviour of Streamwise CVPs at High AoAs

The significant unsteadiness in the streamwise CVPs observed during the experiments provided the authors strong motivations to investigate further. While New et al. (2016) provided an initial look into this particular behaviour, it could not be ascertained whether free-stream turbulence level or mutual interactions contributed more towards the observation. Additionally, it should be noted that streamwise CVPs are inherently transient flow entities formed by separated flows around the LE tubercles. While their influences upon the overall flow behaviour, aerodynamic performances and mean wall shear distribution along hydrofoil suction surface had been studied numerically (Yoon et al. 2011; Favier et al. 2012; Rostamzadeh et al. 2013; Wang et al. 2014; Skillen et al. 2015; Kim et al. 2016; Pérez-Torró and Kim 2017; Serson et al. 2017a, b; Chen et al. 2018), the transient behaviour of the streamwise CVPs have not been looked into more previously. This is especially the case for which the formation of streamwise CVPs begins to initiate once the LE tubercled hydrofoil is subjected to the presence of a free-stream from rest. To address the preceding two issues, a transient numerical study was thus conducted to shed more light upon them. In this case, a LE tubercled NACA634-021 A9λ37.5 hydrofoil inclined at an AoA of $\alpha = 20°$ was used with a free-stream at Re = 14,000. To look into the transient flow features, the effects of the streamwise CVPs upon the wall shear distribution from time t = 0 till they become fully-formed, and the results are presented in Fig. 14. Note that the simulated flow field was tracked up to t = 19 s, where the unsteady

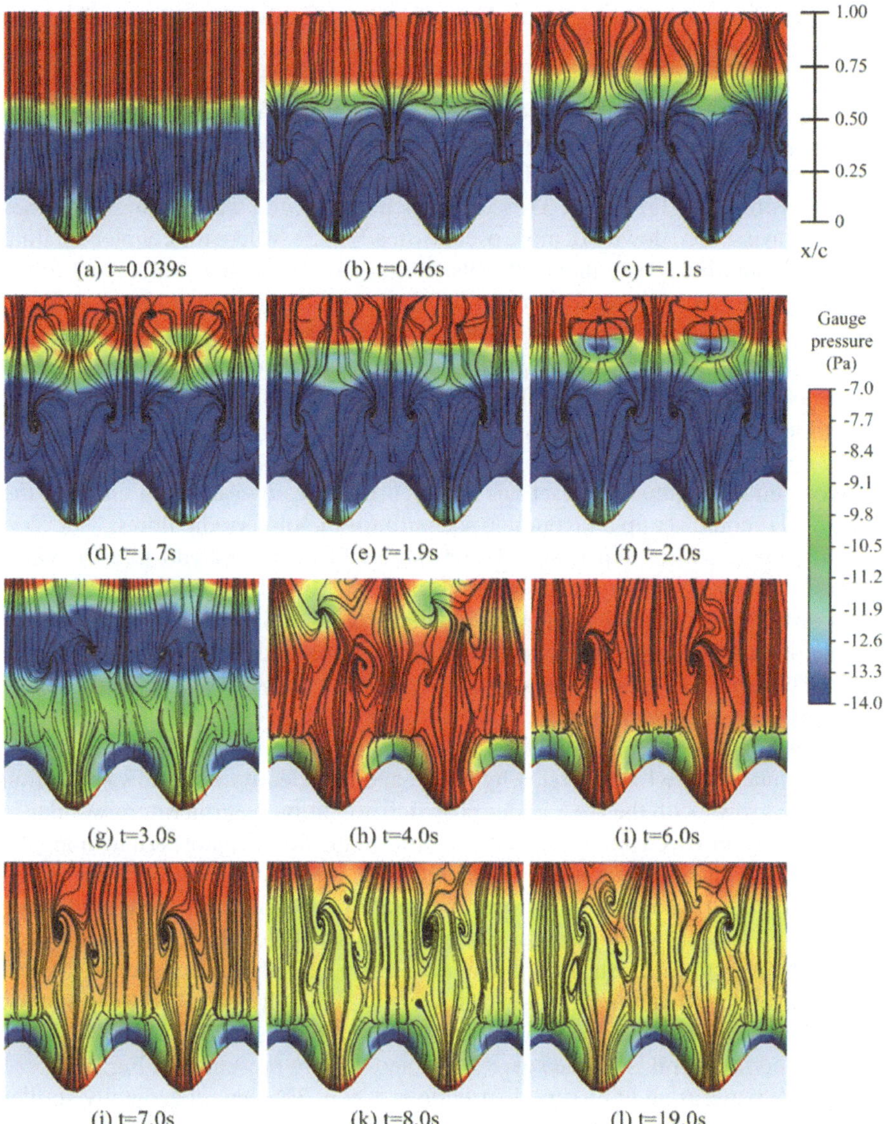

Fig. 14 Temporal evolution of wall shear streamlines along the NACA634-021 A9λ37.5 hydrofoil suction surface at $\alpha = 20°$

flow field results become relatively invariant with computational run time, and the plots comprise of wall shear streamlines and surface pressure distributions.

Shortly after the free-stream was imposed in the simulation, distinct pressure gradients are produced around the LE tubercles and the hydrofoil surface. As the flow progresses from t = 0.46 s onwards, distinctive wall shear streamline features in the form of counter-rotating streamlines aft of the LE tubercles resulting from the formation of the streamwise CVPs can be clearly observed. It should be highlighted that the surface pressure distribution varies as the wall shear streamlines change according to the transient formation of the streamwise CVPs. In particular, those associated with regions downstream of x/c = 0.75 location can be seen to gradually change with time, with transient flow structures and critical points forming close to the hydrofoil TE. From t = 3.0 s onwards however, the wall shear streamlines appear to have settled down to a more pseudo-steady state, where the counter-rotating streamlines, and hence streamwise CVPs, are the dominant flow features up to the mid-chord location. The surface pressure distribution also converges towards a more steady-state appearance from t = 8 s thereafter. It should also be mentioned that the wall shear streamlines at this point resemble the surface oil flow visualization patterns observed by Wei et al. (2016a). Interestingly, the flow visualization results reported in that experimental study revealed the presence of bi-periodic flows close to the TE of the LE tubercled wing. If one inspects Fig. 14k, l closely, it is plausible that more intense mutual interactions between the streamwise CVPs close to the hydrofoil TE could set up a favourable scenario for the bi-periodic flows.

Lastly, numerical or experimental studies on LE tubercled hydrofoils or wings immersed in a free-stream often do not go beyond AoAs that far exceed the stall-angles associated with their particular baseline configurations. However, it is very possible that certain application may incur large AoAs under certain scenarios and it will be insightful to explore the resulting flows when the AoAs far exceeds typical stall-angles. As such, AoAs of $\alpha = 30°$ and $40°$ were investigated and the results are presented in Figs. 15 and 16. Figure 15 shows that streamlines taken along the T- and P-planes of the LE tubercled hydrofoil and it is clear that the flow separation bubble size grows with the AoA under post-stall conditions. Furthermore, similar to what had been observed under pre-stall conditions, the flow separates right along the LE along the T-plane, while it does so at about the mid-chord location along the P-plane. It is interesting to note that the flow separation location along the P-plane does not vary significantly when the AoA increases by $10°$ here. However, it is unclear whether the overall flow behaviour along the LE tubercled hydrofoil suction surface remains invariant when the AoA varies from $\alpha = 30°$ to $40°$. To shed light on this, wall shear streamlines and pressure distributions along the suction surface when the LE tubercled hydrofoil is pitched at $\alpha = 30°$ and $40°$ are presented in Fig. 16.

The flow patterns and pressure distributions at $\alpha = 30°$ remain generally similar to those observed for $\alpha = 20°$ earlier one. At $\alpha = 40°$ however, the flow behaviour becomes discernibly different, where the streamwise CVP vortex filaments appear to interact and merge in a regular fashion from the mid-chord location onwards. In doing so, the wall shear streamlines rearrange into what appears to be bi-periodic flow patterns close to the hydrofoil TE, where the number of CVPs will be a fraction of the LE tubercle numbers. Of particular interest is the concurrent and significant pressure drop around the trough and peak locations, not to mention the overall reduction in suction surface pressure. The realignments of flow patterns along the hydrofoil suction surface also produce critical points that are reminiscent of those observed by Rostamzadeh et al. (2014).

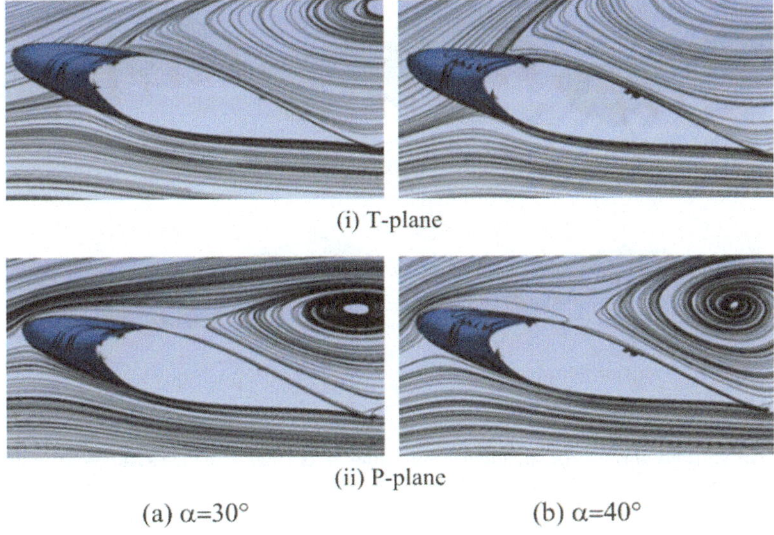

Fig. 15 Velocity streamlines for the LE tubercled hydrofoil at **a** $\alpha = 30°$ and **b** $\alpha = 40°$ along the (i) T- and (ii) P-planes

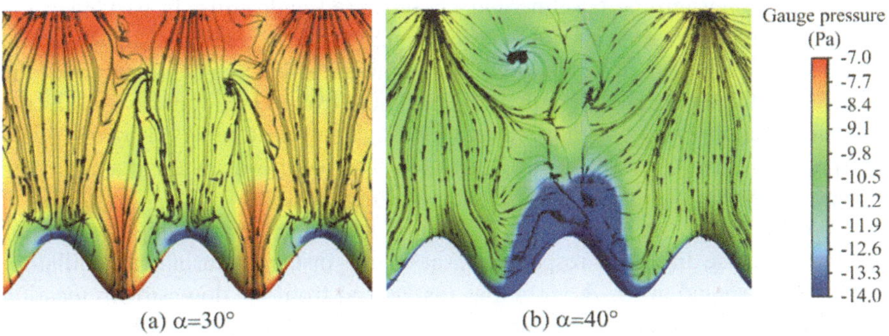

Fig. 16 Wall shear streamlines and pressure distributions along the LE tubercled hydrofoil suction surface at **a** $\alpha = 30°$ and (b) $\alpha = 40°$

5 Oscillating LE Tubercled Hydrofoil

The current study made use of a 10° oscillation amplitude, as well as a reduced frequency (k) of 4.96 with a corresponding Strouhal number (St) of 0.37. Strouhal number is used to compare the current study with previous studies on oscillating smooth (Koochesfahni 1989; Ramamurti and Sandberg 2001; Xiao and Liao 2009). The definition of Strouhal number for oscillating hydrofoils follows that used by Triantafyllou et al. (1991). Based on an earlier study by Benaissa et al. (2017), a static baseline hydrofoil will produce relatively stable flows around the hydrofoil suction

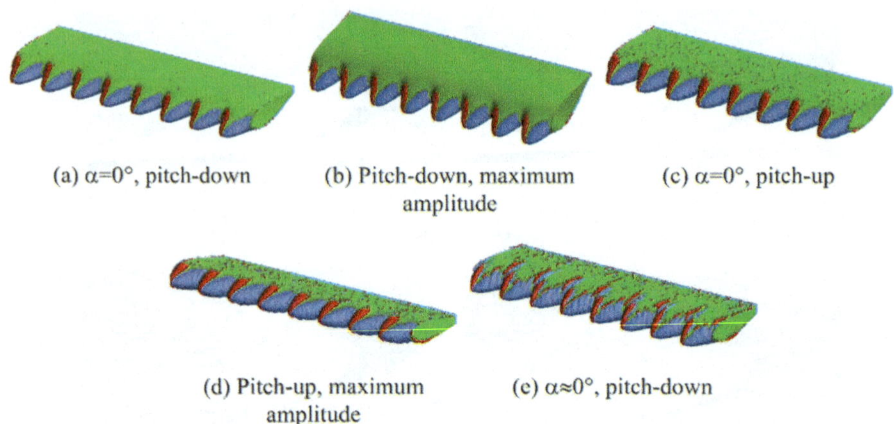

(a) $\alpha=0°$, pitch-down   (b) Pitch-down, maximum amplitude   (c) $\alpha=0°$, pitch-up

(d) Pitch-up, maximum amplitude   (e) $\alpha\approx0°$, pitch-down

Fig. 17 Iso-vorticity surfaces of streamwise vorticities associated with the LE tubercles at different pitching instances

surface at small to moderate AoAs. Under oscillatory conditions, the same is observed whereby the smooth hydrofoil does not exhibit significant complex flow patterns. In additional to the differences in wall shear streamlines, vortex-shedding behavior is also different between the baseline and LE tubercled hydrofoils. In particular, the latter shows that the vortex shedding at the trough location is always slightly ahead of the vortex shed at the peak location. This behavior is due to a combination of the difference in the local chord length at the peak and trough locations, as well as the converging and diverging streamlines at the trough and peak locations respectively (i.e. spanwise flows). However, little attention was paid towards the detailed flow changes associated with the actual LE tubercles in that study.

Figure 17 shows the iso-surfaces of y-vorticity (i.e. x and z being the streamwise and spanwise directions respectively) at various instances during the oscillatory motion. Positive and negative vorticities (as viewed from the downstream location towards the hydrofoil LE) are highlighted in red and blue respectively, while the hydrofoil surface is colored in green. To avoid cluttering the figures, vortices being shed from the hydrofoil TE are not shown in this figure. It can be observed that throughout the pitching cycle, positive and negative vorticities are being produced at the hydrofoil LE. This implies that formations of streamwise CVPs are robust and not adversely affected by the oscillations, at least for the frequency and amplitude used here. However, the streamwise CVPs do not appear to persist very far downstream of the LE tubercles, with the exception of pitch-downstage close to AoA of 0° shown in Fig. 17e, i. Presumably, this is due to the pitch-down motion which enhances the flow separations around the LE tubercles and strengthens their formations and subsequent persistence. Interestingly, closer inspection also reveals that the streamwise CVPs extend further downstream up to mid-chord location and their cores interact with their adjacent neighbors downstream of the trough locations at this moment. For these observations, it can be deduced that at least some aspects of the streamwise

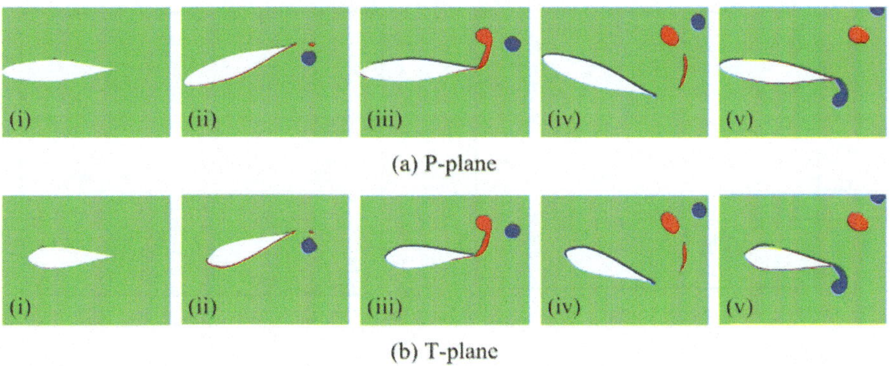

(a) P-plane

(b) T-plane

Fig. 18 Vortices shed from the **a** P- and **b** T-planes at pitching instances corresponding to **a** to **e** in Fig. 17

CVP behavior are dynamic. Furthermore, from the side-view z-vorticity contours shown in Fig. 18, it can be observed that no significant flow separation occurs along the hydrofoil surface throughout the entire oscillation cycle as well. In particular, it should be noted that the vortices shed from the TE of the LE tubercled hydrofoil do not incur any discernible phase lag between the P and T-planes.

Lastly, thrust and power coefficients, as well as propulsive efficiency are plotted against Strouhal number in Fig. 19 and compared to the previous studies on non-tubercled hydrofoils (Koochesfahni 1989; Ramamurti and Sandberg 2001; Xiao and Liao 2009) undergoing oscillatory motions. Note that the comparison is based on similar Strouhal numbers across all these studies and should give some initial appreciation on any possible thrust generation. Note that the Strouhal number here is defined as $St = (3k\sin(A))/2\pi$ for the sake of consistency across the different studies. On the other hand, thrust coefficient is defined as the negative of the drag coefficient, while power coefficient is defined as $C_{ip} = (M\dot{\theta})/(1/2\rho U^3 c)$, where M is the pitching moment per meter span, $\dot{\theta}$ is the rotation rate and ρ is the fluid medium density. With these definitions, propulsive efficiency can then be further defined as $\eta = C_T/C_{ip}$. Results show that the LE tubercled hydrofoil studied here produces a higher thrust coefficient but lower power coefficient at the present operating point. This leads to a slightly higher propulsive efficiency as compared to the other studies. It is also worth noticing that while the efficiency level for conventional non-tubercled hydrofoils decreases after $St = 0.15$, that for the LE tubercled hydrofoil remains significantly higher at $St = 0.4$. This suggests that it may be beneficial to explore LE tubercled hydrofoils further for use as propulsive devices under oscillatory conditions.

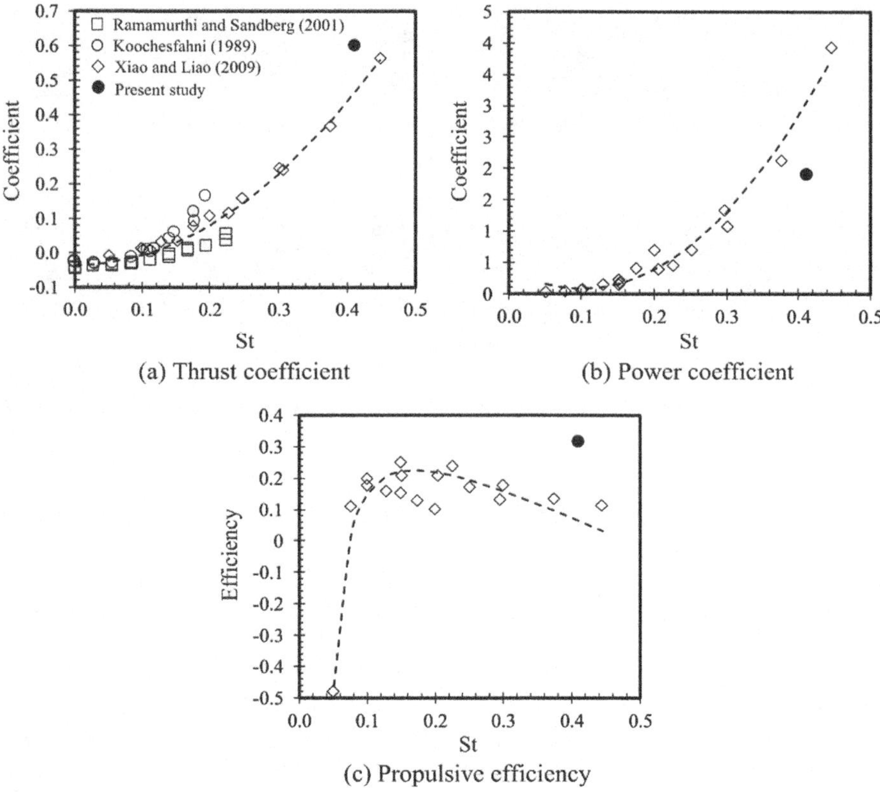

(a) Thrust coefficient

(b) Power coefficient

(c) Propulsive efficiency

Fig. 19 Comparisons of **a**, **b** thrust and power coefficients, as well as **c** efficiency against results from earlier experimental and numerical studies

6 Conclusions

The present chapter covers some of the more recent experimental and numerical developments in the implementation of LE tubercles in hydrofoils. Firstly, the use of LE tubercled hydrofoils under low Reynolds number conditions where hydrofoil profiles are cambered rather than the more common non-cambered ones was explored, where results indicate that LE tubercles remain effective in mitigating flow separation and stall despite the presence of significant adverse pressure gradients. Secondly, transient developments of streamwise CVPs produced by LE tubercles have been investigated and numerical results at high AoAs suggest that reorganization of the streamwise CVPs are responsible for the bi-periodic flow behaviour along the hydrofoil TE. Last but not least, the flow behaviour and possible thrust production by an oscillating LE tubercled hydrofoil has been explored, where preliminary results suggest that its propulsive efficiency may exceed those produced by non-tubercled hydrofoils.

Acknowledgements The authors gratefully acknowledge the support provided by Singapore Ministry of Defense—Future Systems and Technology Directorate Defense Innovative Research Programme, Nanyang Technological University, Temasek Laboratories @ National University of Singapore, National Natural Science Foundation of China (grant no. 11702173 and 41527901), and South Africa National Research Foundation CSUR (grant no. 98876).

References

Benaissa M, Ibrahim IH, New TH, Ho WH (2017) Effect of leading edge protuberances on thrust production of a dynamically pitching hydrofoil. In: Conference proceedings of topical problems of fluid mechanics, pp. 23–30. Prague

Bolzon MD, Kelso RM, Arjomandi M (2016) Formation of vortices on a tubercled wing, and their effects on drag. Aerosp Sci Technol 56:46–55

Chaitanya P, Joseph P, Narayanan S, Vanderwel C, Turner J, Kim JW, Ganapathisubramani B (2017) Performance and mechanism of sinusoidal leading edge serrations for the reduction of turbulence–aerofoil interaction noise. J Fluid Mech 818:435–464

Chen H, Wang JJ (2014) Vortex structures for flow over a delta wing with sinusoidal leading edge. Exp Fluids 55(6):1761

Chen W, Qiao W, Tong F, Wang L, Wang X (2018). Numerical investigation of wavy leading edges on rod–airfoil interaction noise. AIAA J 1–15

Custodio D (2007) The effect of humpback whale-like leading edge protuberances on hydrofoil performance. Worcester Polytechnic Institute, MA

Custodio D, Henoch CW, Johari H (2015) Aerodynamic characteristics of finite span wings with leading-edge protuberances. AIAA J 53(7):1878–1893

Favier J, Pinelli A, Piomelli U (2012) Control of the separated flow around an airfoil using a wavy leading edge inspired by humpback whale flippers. CR Mec 340(1–2):107–114

Fish FE, Battle JM (1995) Hydrodynamic design of the humpback whale flipper. J Morphol 225(1):51–60

Fish FE, Lauder GV (2006) Passive and active flow control by swimming fishes and mammals. Annu Rev Fluid Mech 38:193–224

Fish FE, Howle LE, Murray MM (2008) Hydrodynamic flow control in marine mammals. Integr Comp Biol 48(6):788–800

Fish FE, Weber PW, Murray MM, Howle LE (2011a) The tubercles on humpback whales' flippers: application of bio-inspired technology. Integr Comp Biol 51(1):203

Fish FE, Weber PW, Murray MM, Howle LE (2011b) Marine applications of the biomimetic Humpback whale flipper. Marine Technol Soc J 45(4):198–207

Flint TJ, Jermy MC, New TH, Ho WH (2017) Computational study of a pitching bio-inspired corrugated airfoil. Int J Heat Fluid Flow 65:328–341

Goruney T, Rockwell D (2009) Flow past a delta wing with a sinusoidal leading edge: near-surface topology and flow structure. Exp Fluids 47(2):321–331

Guerreiro JLE, Sousa JMM (2012) Low-Reynolds-number effects in passive stall control using sinusoidal leading edges. AIAA J 50(2):461–469

Hansen KL, Kelso RM, Dally BB (2011) Performance variations of leading-edge tubercles for distinct airfoil profiles. AIAA J 49(1):185

Hansen KL, Rostamzadeh N, Kelso RM, Dally BB (2016) Evolution of the streamwise vortices generated between leading edge tubercles. J Fluid Mech 788:730–766

Ibrahim IH, New TH (2015) Tubercle modifications in marine propeller. In: Conference proceedings of 10th Pacific symposium on flow visualization and image processing, Naples, Italy

Johari H (2015) Cavitation on hydrofoils with sinusoidal leading edge. J Phys: Conf Ser 656(1):012155

Johari H, Henoch CW, Custodio D, Levshin A (2007) Effects of leading-edge protuberances on airfoil performance. AIAA J 45(11):2634–2642

Joy J, Ibrahim IH, New TH (2016) Numerical investigation on flow separation control of low Reynolds number sinusoidal aerofoils. In: Proceedings of 46th AIAA fluid dynamics conference, 3949

Keane RD, Adrian RJ (1992) Theory of cross-correlation analysis of PIV images. Appl Sci Res 49(3):191–215

Kim JW, Haeri S, Joseph PF (2016) On the reduction of aerofoil–turbulence interaction noise associated with wavy leading edges. J Fluid Mech 792:526–552

Koochesfahni MM (1989) Vortical patterns in the wake of an oscillating airfoil. AIAA J 27(9):1200–1205

Lin JC (2002) Review of research on low-profile vortex generators to control boundary-layer separation. Prog Aerosp Sci 38(4):389–420

Lohry M, Clifton D, Martinelli L (2012) Characterization and design of tubercle leading-edge wings. In: Conference proceedings of computational fluid dynamics (ICCFD7), Big Island, Hawaii

Miklosovic DS, Murray MM, Howle LE, Fish FE (2004) Leading-edge tubercles delay stall on humpback whale (*Megaptera novaeangliae*) flippers. Phys Fluids 16(5):L39–L42

Miklosovic DS, Murray MM, Howle LE (2007) Experimental evaluation of sinusoidal leading edges. J Aircraft 44(4):1404

Moffat RJ (1988) Describing the uncertainties in experimental results. Exp Thermal Fluid Sci 1(1):3–17

Murray MM, Miklosovic DS, Fish FE, Howle LE (2005) Effects of leading edge tubercles on a representative whale flipper model at various sweep angles. In: Conference proceedings of the 14th international symposium on unmanned untethered submersible technology, Durham, NH

New TH, Wei ZY, Koh GC, Cui YD (2014) On the role and effectiveness of streamwise vortices in leading-edge modified wings at low Reynolds numbers. In: Conference proceedings of 16th international symposium on flow visualization, Okinawa, Japan

New TH, Wei ZY, Cui YD (2015) On the effectiveness of leading-edge modifications upon cambered SD7032 wings. In: Conference proceedings of 10th Pacific symposium on flow visualization and image processing, Naples, Italy

New TH, Wei Z, Cui YD (2016) A time-resolved particle-image velocimetry study on the unsteady flow behaviour of leading-edge tubercled hydrofoils. In: Conference proceedings of 18th international symposium on applications of laser and imaging techniques to fluid mechanics, Lisbon, Portugal

Pérez-Torró R, Kim JW (2017) A large-eddy simulation on a deep-stalled aerofoil with a wavy leading edge. J Fluid Mech 813:23–52

Ramamurti R, Sandberg W (2001) Simulation of flow about flapping airfoils using finite element incompressible flow solver. AIAA J 39(2):253–260

Rostamzadeh N, Kelso RM, Dally BB, Hansen KL (2013) The effect of undulating leading-edge modifications on NACA 0021 airfoil characteristics. Phys Fluids 25(11):117101

Rostamzadeh N, Hansen KL, Kelso RM, Dally BB (2014) The formation mechanism and impact of streamwise vortices on NACA 0021 airfoil's performance with undulating leading edge modification. Phys Fluids 26(10):107101

Serson D, Meneghini JR, Sherwin SJ (2017a) Direct numerical simulations of the flow around wings with spanwise waviness. J Fluid Mech 826:714–731

Serson D, Meneghini JR, Sherwin SJ (2017b) Direct numerical simulations of the flow around wings with spanwise waviness at a very low Reynolds number. Comput Fluids 146:117–124

Shi W, Rosli R, Atlar M, Norman R, Wang D, Yang W (2016) Hydrodynamic performance evaluation of a tidal turbine with leading-edge tubercles. Ocean Eng 117:246–253

Skillen A, Revell A, Pinelli A, Piomelli U, Favier J (2015) Flow over a wing with leading-edge undulations. AIAA J 53:464–472

Stanway MJ (2008) Hydrodynamic effects of leading-edge tubercles on control surfaces and in flapping foil propulsion. Doctoral dissertation, Massachusetts Institute of Technology

Taylor GK, Nudds RL, Thomas AR (2003) Flying and swimming animals cruise at a strouhal number tuned for high power efficiency. Nature 425:707–711

Triantafyllou MS, Triantafyllou GS, Gopalkrishnan R (1991) Wake mechanics for thrust generation in oscillating foils. Phys Fluids 3(12):2835–2837

Van Nierop EA, Alben S, Brenner MP (2008) How bumps on whale flippers delay stall: an aerodynamic model. Phys Rev Lett 100(5):054502

Wang YY, Hu WR, Zhang SD (2014) Performance of the bio-inspired leading edge protuberances on a static wing and a pitching wing. J Hydrodyn 26(6):912–920

Weber PW, Howle LE, Murray MM (2010) Lift, drag, and cavitation onset on rudders with leading-edge tubercles. Marine Technology 47(1):27–36

Wei, Z., New, T. H., & Cui, Y. D. (2015a). Flow separation analysis of aerofoil with leading-edge tubercles by proper orthogonal decomposition technique. In: Conference proceedings of 13th international conference on fluid control, measurements and visualization (FLUCOME 2015), Doha, Qatar

Wei Z, New TH, Cui YD (2015b) An experimental study on flow separation control of hydrofoils with leading-edge tubercles at low Reynolds number. Ocean Eng 108:336–349

Wei Z, New TH, Cui YD (2016a) Surface flow topology visualizations of wings with leading-edge tubercles under pitch, yaw and roll conditions. In: 17th International symposium on flow visualization, Gatlinburg, Tennessee, USA

Wei Z, Zang B, New TH, Cui YD (2016b) A proper orthogonal decomposition study on the unsteady flow behaviour of a hydrofoil with leading-edge tubercles. Ocean Eng 121:356–368

Wei Z, New TH, Cui YD (2017a) A PIV-based proper orthogonal decomposition study on the vortex structures of tubercled hydrofoils. In: Conference proceedings of 12th international symposium on particle-image velocimetry, Busan, South Korea

Wei Z, New TH, Cui YD (2017b) Aerodynamic performance and surface flow structures of leading-edge tubercled tapered swept-back wings. AIAA J 423–431

Wei Z, Lian L, Zhong Y (2018) Enhancing the hydrodynamic performance of a tapered swept-back wing through leading-edge tubercles. Exp Fluids 59(6):103

Xiao Q, Liao W (2009) Numerical study of asymmetric effect on a pitching foil. Int J Mod Phys C 20(10):1663–1680

Yoon HS, Hung PA, Jung JH, Kim MC (2011) Effect of the wavy leading edge on hydrodynamic characteristics for flow around low aspect ratio wing. Comput Fluids 49(1):276–289

Zhang MM, Wang GF, Xu JZ (2013) Aerodynamic control of low-Reynolds-number airfoil with leading-edge protuberances. AIAA J 51(8):1960–1971

Zhang MM, Wang GF, Xu JZ (2014) Experimental study of flow separation control on a low-Re airfoil using leading-edge protuberance method. Exp Fluids 55(4):1710

Leading-Edge Tubercles on Swept and Delta Wing Configurations

Lihao Feng and Jinjun Wang

Abstract The effects of leading-edge (LE) tubercles on the aerodynamic characteristics of the flipper, swept and delta wings are introduced here. Originated from the whale flipper, it has been demonstrated that LE tubercles could improve the aerodynamic performance of the flipper wing at high angles-of-attack. For the delta wing, the post-stall lift coefficient could be increased by the LE tubercles, while the drag coefficient is also increased, leading to a decrease in the lift-to-drag ratio. Opposite effects are observed on the swept wing due to the difference in test conditions. The study of control mechanism indicates that LE tubercles work as vortex generators to induce streamwise vortices, which are beneficial for flow separation control. Furthermore, the optimal parameters for the tubercles for flow control might differ at different conditions.

Keywords Leading-edge tubercles · Spanwise flow · Flipper wing · Swept wing · Delta wing · Aerodynamic characteristics

1 Background

Leading-edge (LE) tubercles have proven their effects to improve the aerodynamic performances of straight wings at high angles-of-attack (AoAs), which mainly endure streamwise oncoming flows. On the other hand, swept and delta wings, which are fundamental configurations used in modern civil aircrafts and fighters, respectively, endure spanwise flows. Spanwise flow leads to the formation of tip vortex for swept wings and LE vortex for delta wings. As such, it is interesting and important to understand the effects of LE tubercles on swept and delta wings, as the control mechanism is different between the two.

The tip vortex and the LE vortex play important roles on the aerodynamic characteristics of swept and delta wings, respectively. Thus, one of the basic control strategies is to manipulate these vortices to improve the underlying aerodynamics. Tip vortex contributes significantly towards drag penalty, which has an adverse impact

L. Feng · J. Wang (✉)
Institute of Fluid Mechanics, Beihang University, Beijing 100191, China
e-mail: jjwang@buaa.edu.cn

© Springer Nature Switzerland AG 2020
D. T. H. New and B. F. Ng (eds.), *Flow Control Through Bio-inspired Leading-Edge Tubercles*, https://doi.org/10.1007/978-3-030-23792-9_5

on aircraft performances (Spalart 1998; Birch et al. 2004) and a way to overcome it is by weakening the strength of the tip vortex. On the other hand, the LE vortex provides lift for the delta wing (Gursul et al. 2007; Mitchella and Délery 2011). While higher strength of the LE vortex could improve lift coefficient, its breakdown may lead to a sudden drop in lift and hence, it is essential to strengthen this LE vortex in delta wings. Several techniques have been adopted for flow control and most are active methods, such as blowing/suction, synthetic jet, plasma actuator, etc. In comparison, the LE tubercle is a recently developed bio-inspired flow control technique and related studies are still limited.

In this chapter, we present recent progress on the aerodynamics of flipper, swept and delta wings with LE tubercles to understand the control effects, control mechanism, and influence of geometric parameters.

2 Flipper Wing

One of the earliest works to study the aerodynamic characteristics of LE tubercles was conducted by Miklosovic et al. (2004). They used the force balance to measure forces on flipper wings with smooth and LE tubercles, as shown in Fig. 1. The Reynolds number was in the range of 5.05×10^5 to 5.20×10^5. The standard deviations of the measured lift coefficient, drag coefficient, and pitching moment coefficient were ± 0.0003, ± 0.003, and ± 0.001, respectively.

The aerodynamic forces are compared between the smooth and tubercled flipper wings in Fig. 2. It is observed that the lift coefficient for the tubercled wing is nearly the same as the smooth wing for $\alpha \leq 8.5°$. However, the presence of LE tubercles could increase the post-stall lift coefficient of the smooth wing, with the maximum lift

Fig. 1 Flipper wings with **a** smooth and **b** tubercled LEs (Miklosovic et al. 2004)

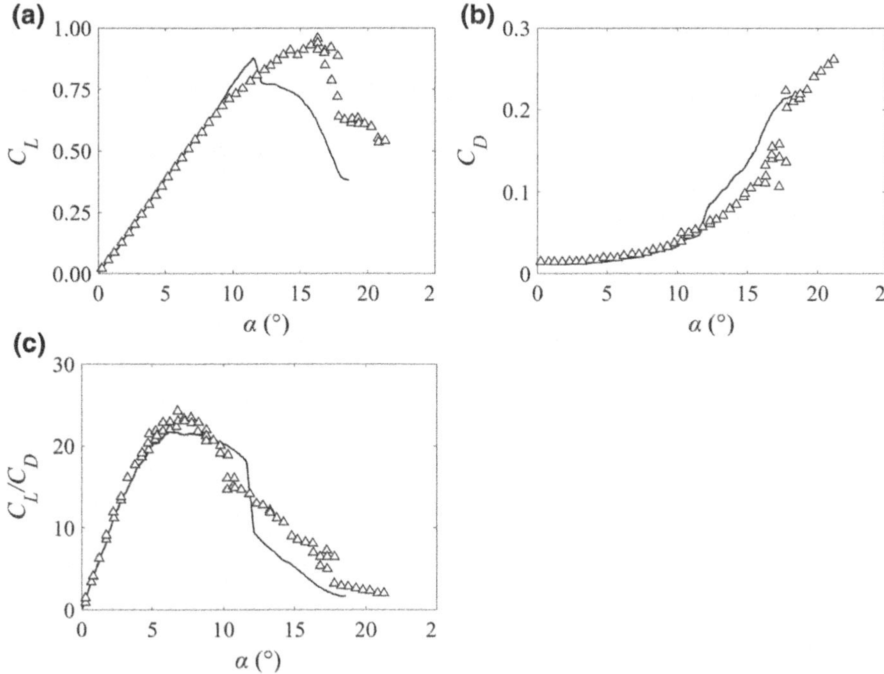

Fig. 2 Aerodynamic forces of the smooth and tubercled whale flipper wings. **a** Lift coefficient; **b** drag coefficient; **c** lift-to-drag ratio. The solid line and the discrete triangles denote the smooth and tubercled cases, respectively (Miklosovic et al. 2004)

coefficient increased by about 6%. The stall angle is also delayed from the original $\alpha = 12°$ to about $\alpha = 16°$. As compared to the smooth wing, drag coefficient is reduced by the presence of the LE tubercles, particularly in the range of $12° < \alpha < 19°$. Consequently, the lift-to-drag ratio is increased for nearly the entire range of AoAs except for $9° < \alpha < 12°$. The maximum lift-to-drag ratio is increased by about 13%.

Subsequent works, such as Miklosovic et al. (2007), van Nierop et al. (2008), and Custodio et al. (2015) further investigated the effects of LE tubercles. It is observed that the aerodynamic performance of the tubercled flipper wing is improved at high AoAs, which explains well the existence of the LE tubercles for the humpback whale flipper.

The variation of the flow over the flipper suction surface induced by LE tubercles is shown in Fig. 3. At $\alpha = 8°$, flow separation begins to form near the tip of the smooth flipper, indicating the formation of the tip vortex which can be eliminated by the LE tubercles. At $\alpha = 12°$, the flow separation region at the tip of the smooth flipper grows larger and trailing-edge (TE) stall develops along most of the TE of the flipper. A distinct stall line could be observed in the middle portion of the smooth flipper, from about 20 to 80%. When the AoA further increases to $\alpha = 16°$, which is in the

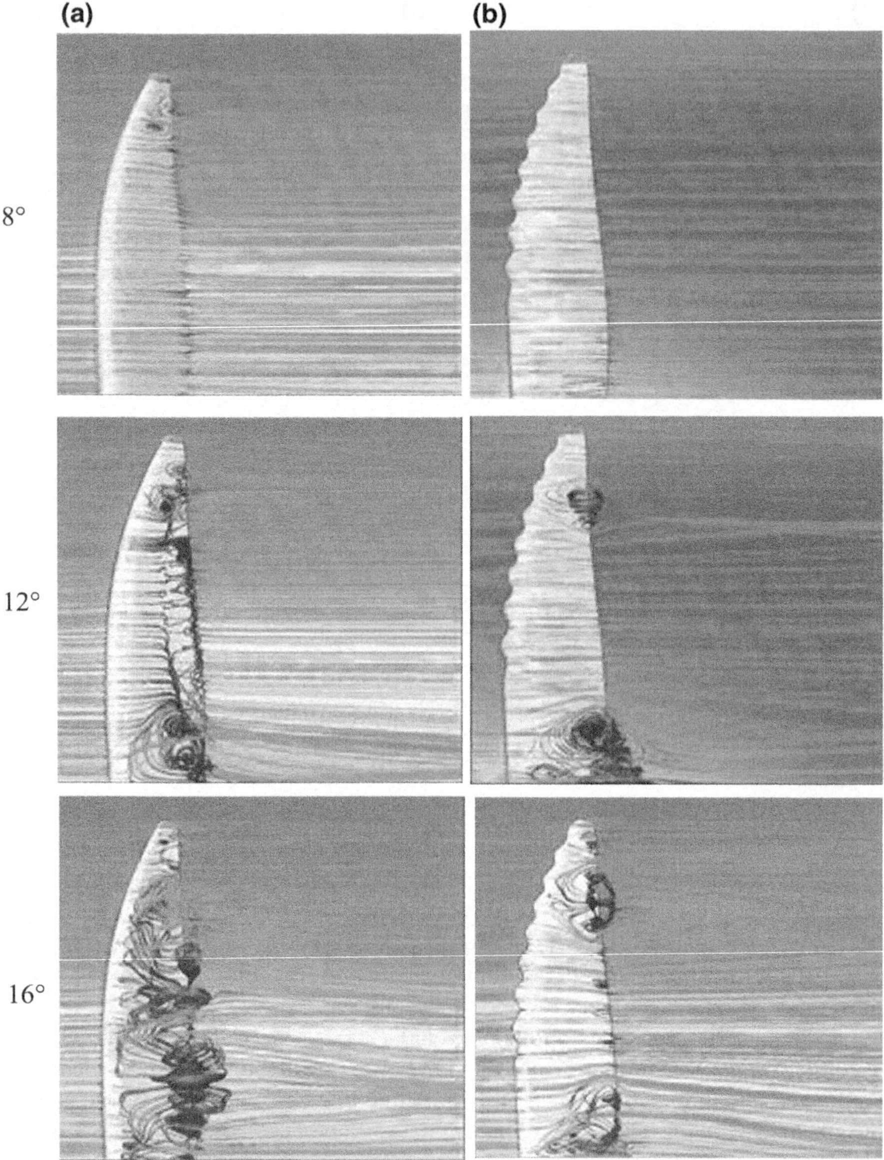

Fig. 3 Streamlines over the surface of the **a** smooth and **b** tubercled whale flipper wings at different AoA of $\alpha = 8°$, 12°, and 16° (Weber et al. 2011)

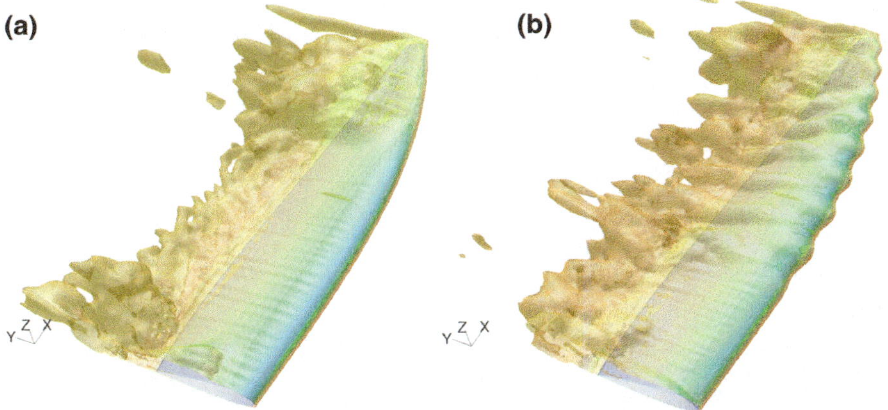

Fig. 4 Instantaneous vorticity magnitude isosurface of the **a** smooth and **b** tubercled whale flipper wings at $\alpha = 15°$ (Pedro and Kobayashi 2008)

stall region, a large flow separation region develops across the entire surface of the smooth flipper. In comparison, the LE tubercles completely eliminate the separation region in the middle portion for both $\alpha = 12°$ and $16°$.

The instantaneous vorticity magnitude isosurface is shown in Fig. 4. It indicates that both smooth and tubercled flipper wings produce similar vortical structures at the root and the tip. However, the tubercled flipper displays streamwise vortices aligned with the tubercles along the flipper midsection. In comparison, the smooth flipper has no such flow structures. From the slices of the vorticity magnitude in Fig. 5, differences in the vorticity fields can be observed where high values of streamwise vorticity are located in the middle portion of the tubercles flipper.

From Figs. 4 and 5, it is evident that LE tubercles function as vortex generators to induce streamwise vortices. The streamwise vortical structures are beneficial to enhance momentum mixing of the surrounding fluids, thus enabling the flow to overcome adverse pressure gradients for flow separation delay or even separation elimination. This can be validated by the instantaneous $|V|^2$ isosurface shown in Fig. 6. The streamwise vortices carry high momentum flow close to the wall to reenergize the boundary layer. Thus, the area of $|V|^2$ isosurface is enlarged by the LE tubercles in the middle portion of the flipper wing.

3 Swept Wing

It is only recently that LE tubercles are proposed to be used on swept wings. As such, this section will document preliminary results from two groups with different approaches, which serves as a reference for future investigations.

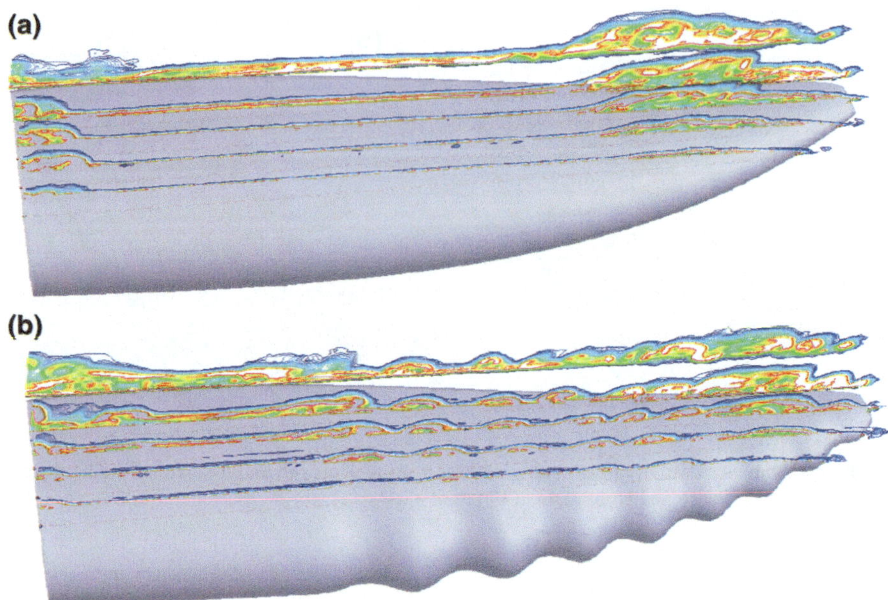

Fig. 5 Instantaneous vorticity magnitude slices at different streamwise directions for the **a** smooth and **b** tubercled whale flipper wings at $\alpha = 15°$ (Pedro and Kobayashi 2008)

Bolzon et al. (2016, 2017a, b) conducted a series of studies of the tubercled swept wing. Two swept wing models with smooth and tubercled LE were investigated, and the latter is shown in Fig. 7a. For the smooth wing, it had a span of 330 mm, a root chord of 175 mm, a wingtip chord of 70 mm, and a mean aerodynamic chord of 130 mm. The sweep angle of the wing was set to 35° and the taper ratio was 0.4. The tubercles were constructed with a constant thickness-to-chord ratio. They had a constant amplitude of 10.5 mm and a constant wavelength of 60 mm. Experiments were carried out in the KCWind Tunnel at the University of Adelaide, as shown in Fig. 7b. The free-stream velocity was at $U_\infty = 27.5$ m/s and the turbulence intensity was 0.6–0.8%, corresponding to a Reynolds number of $Re = 2.25 \times 10^5$. The aerodynamic forces on the wings were measured by a six-component JR3 brand load cell. The load cell had a 95% confidence interval of $\pm1.75\%$ for a force of 0.5 N to $\pm1\%$ for a force of 2.2 N in the x-direction, and the y-direction corresponded to $\pm1.76\%$ for a force of 0.5 N to $\pm1.5\%$ for a force of 5.9 N.

The lift and drag coefficients for both the smooth and tubercled wings are shown in Fig. 8a, and the lift-to-drag ratio in Fig. 8b. For $0 \leq \alpha \leq 8°$, the LE tubercles reduced the lift coefficient by about 4–6% and the drag coefficient by about 7–9.5%, resulting in an increase in the lift-to-drag ratio of about 2–6%. The maximum lift-to-drag ratio at $\alpha = 6°$ is also increased by about 3%. However, for $\alpha > 8°$, the tubercles reduced the lift coefficient while increasing the drag coefficient, resulting in a reduced lift-to-drag ratio. Bolzon et al. (2017a) suggested that this sudden drop in performance

Fig. 6 Instantaneous $|V|^2$ isosurface of the **a** smooth and **b** tubercled whale flipper wings at $\alpha = 15°$ (Pedro and Kobayashi 2008)

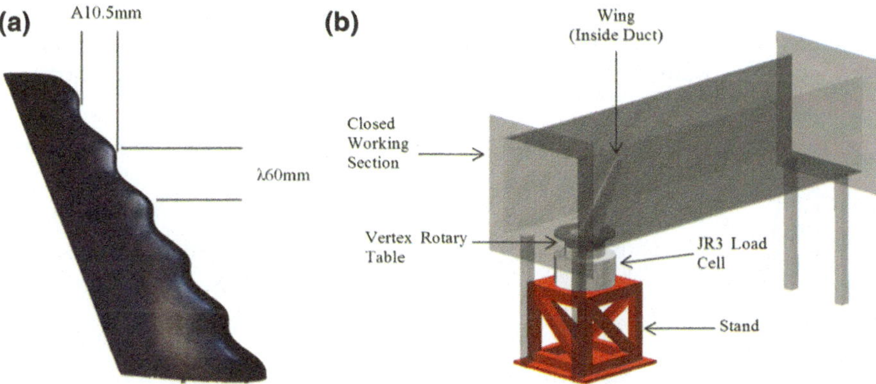

Fig. 7 **a** Swept wing model with LE tubercles; **b** experimental setup for the force measurement of the swept wing in a wind tunnel (Bolzon et al. 2017a)

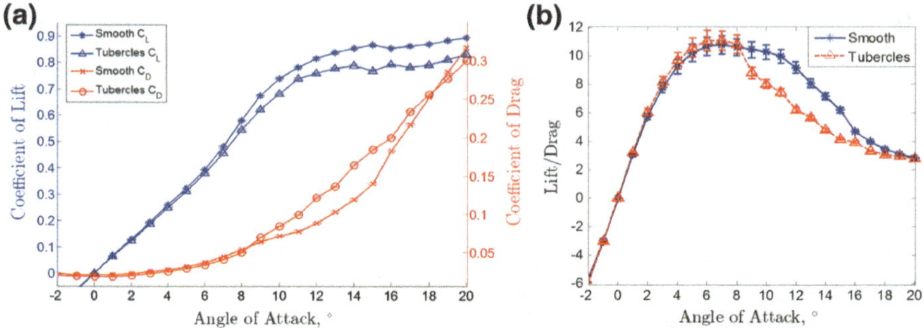

Fig. 8 Aerodynamic forces of the smooth and tubercled swept wing. **a** Lift and drag coefficients; **b** lift-to-drag ratio (Bolzon et al. 2017a)

of the tubercled wing was due to the premature onset of flow separation. They found that while tubercles typically softened the stall for un-swept wings, they had little effect on the severity of the stall region for this swept wing.

Wei et al. (2018) conducted experiments on the aerodynamic characteristics of the tubercled swept wing. The swept wing models and the experimental setup are shown in Fig. 9. The smooth wing was based on a SD7032 airfoil along the streamwise cross-sections, which had high aerodynamic performance at low Reynolds numbers. The root chord of the model was 112.5 mm, the tip chord was 37.5 mm and the span was 300 mm, resulting in a taper ratio of 0.33. The sweep angle based on the quarter-chord line was 30°, and the corresponding mean aerodynamic chord was 81.25 mm. For the tubercled wing, the amplitude was kept at 0.12 of the local chord c. The wavelength was kept constant at 0.47 c. The platform areas for the smooth and tubercled wings were 0.0225 and 0.0226 m^2, respectively, while the aerodynamic

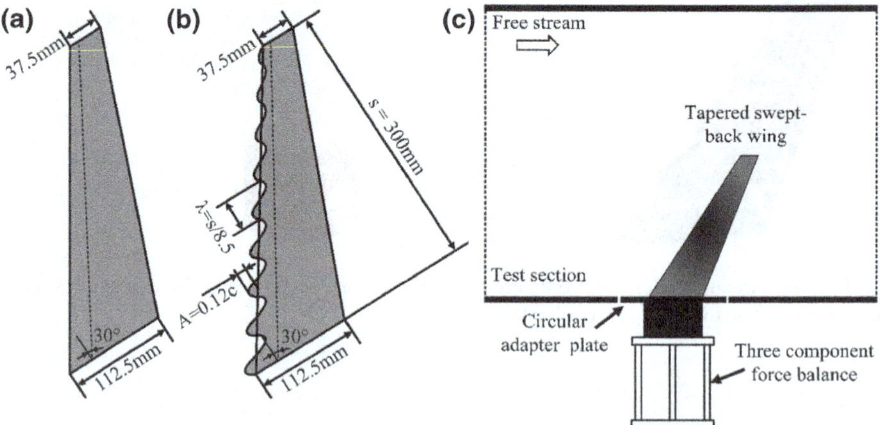

Fig. 9 **a** Smooth and **b** tubercled swept wing models; **c** experimental setup for the force measurement of the swept wing in a wind tunnel (Wei et al. 2018)

forces were measured by a three-component force balance in a closed-loop wind tunnel. The free-stream velocity was set at $U_\infty = 10$ m/s, corresponding to the Reynolds number of $Re = 5.5 \times 10^4$. The turbulent intensity in the test section was less than 0.25%, and the flow uniformity was below 0.5%. The total uncertainty for the measured force was estimated to be approximately 2.7%.

The force coefficients are shown in Fig. 10 for both smooth and tubercled cases. They indicated that the lift coefficient for the tubercled wing is enhanced at all the AoAs tested, even though the maximum lift coefficient will not increase as significantly. The tubercled wing produces slightly higher drag coefficients than the smooth wing for $10° \le \alpha \le 20°$, which covers both the pre- and post-stall regimes. Consequently, the LE tubercles could increase the lift-to-drag ratio up to $\alpha = 10°$. The maximum lift-to-drag ratio of 6.32 occurs at $\alpha = 4°$, which is approximately 25% higher than that for the smooth wing.

Figure 11 shows the oil flow visualization over the suction surface of the smooth wing at five typical AoAs, and Fig. 12 shows the flow features interpreted from these images. At $\alpha = 0°$, a laminar separation bubble forms over the suction surface and extends along the entire wing span. As the AoA increases to $10°$ and $15°$, the laminar

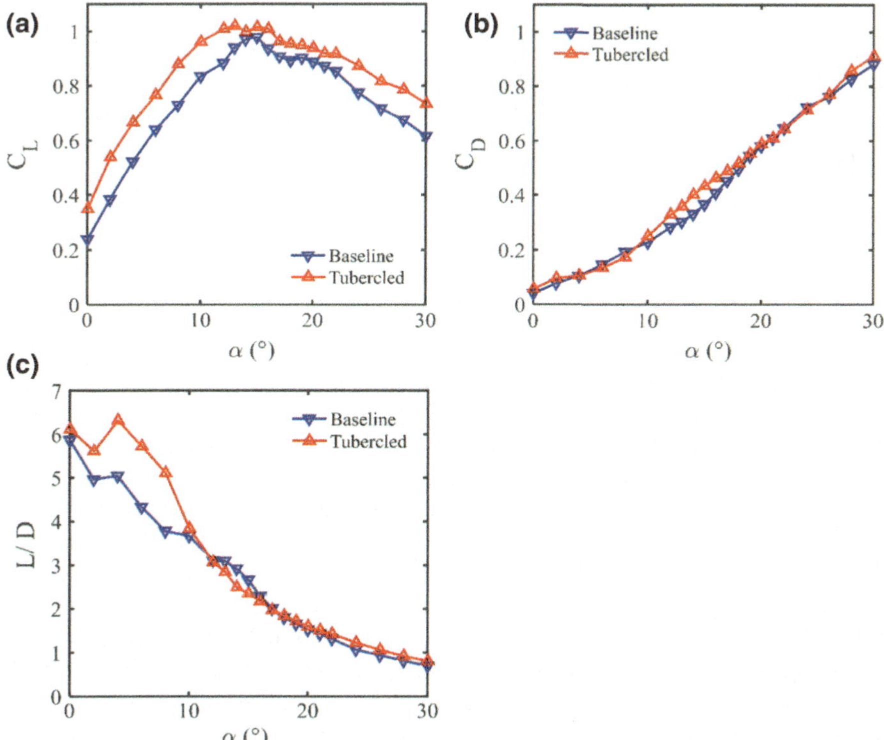

Fig. 10 Aerodynamic forces of the smooth and tubercled swept wings. **a** Lift coefficient; **b** drag coefficient; **c** lift-to-drag ratio (Wei et al. 2018)

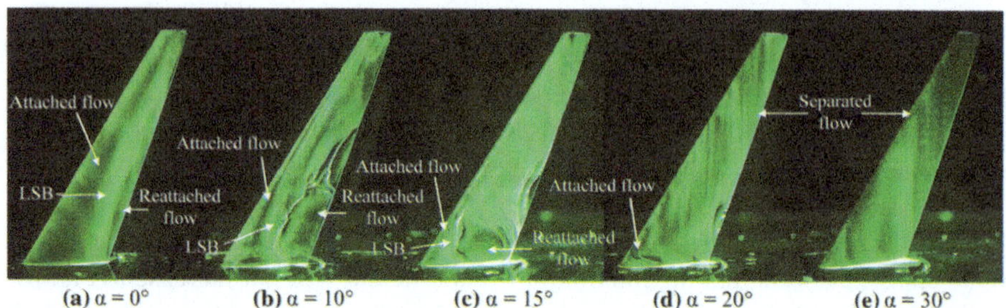

Fig. 11 Oil flow visualization over the suction surface of the smooth wing (Wei et al. 2018)

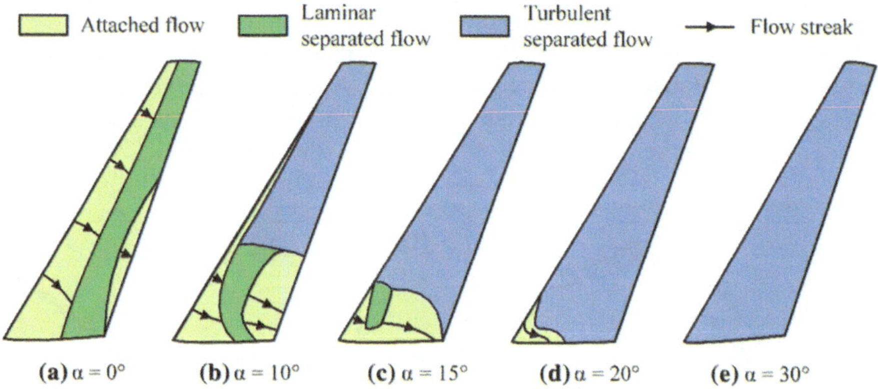

Fig. 12 Sketches of the flow features interpreted from the oil flow visualization of the smooth wing (Wei et al. 2018)

separation bubble gradually decreases to a short spanwise one, which is located in the inboard region close to the wing root. After $\alpha > 15°$, the laminar separation bubble disappears and turbulent flow separation occurs from the LE, with almost no existence of attached flow.

Figures 13 and 14 show the flow visualization and the corresponding interpreted sketches for the tubercled wing, respectively. At $\alpha = 0°$, the laminar separation bubble can still be observed. However, it is distributed in a wave-like pattern, which is segregated by the small streamwise vortices downstream of the tubercle troughs. When the AoA increases to $10° \leq \alpha \leq 20°$, the laminar separation bubble decreases its size near the wing root. Turbulent flow separation occurs in the regions downstream of the tubercle troughs, while the flow remains attached in the downstream regions close to the tubercle crests. In comparison with the smooth case, the tubercled wing exhibits more areas of attached flow, even at $\alpha = 30°$. This observation clearly explains why the tubercled wing could increase the lift coefficient of the swept wing for the entire AoAs.

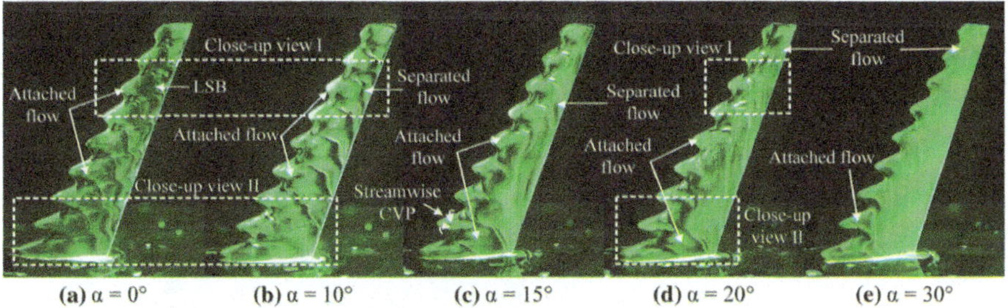

Fig. 13 Oil flow visualization over the suction surface of the tubercled wing (Wei et al. 2018)

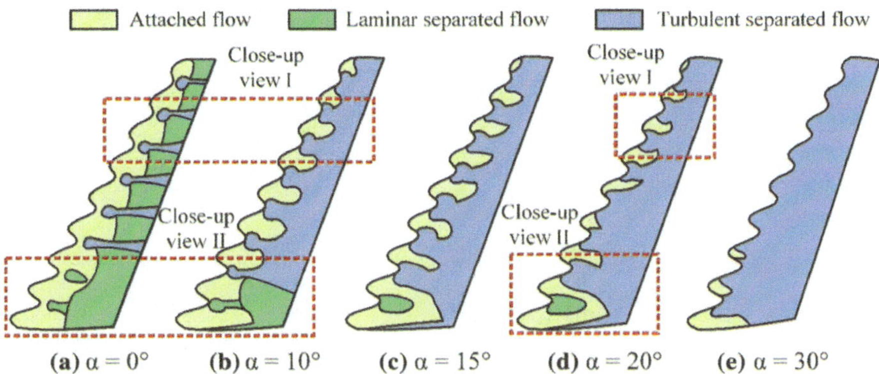

Fig. 14 Sketches of the flow features interpreted from the oil flow visualization of the tubercled wing (Wei et al. 2018)

4 Delta Wing

Chen et al. (2013) and Chen and Wang (2014) conducted a series of investigations into the effects of LE tubercles on a delta wing. Both smooth and tubercled models were studied. The smooth delta wing model was a flat plate with a sweep angle of 52° without beveling on either side of the wing. The root chord length was 200 mm, and the wing thickness was 3 mm. For the tubercled wing, sinusoidal curves with troughs and crests were imposed along the LE, as shown in Fig. 15. In order to study the effects of the amplitude (A) and wavelength (λ) of the tubercles on the aerodynamic performances of the delta wing, different parameters were chosen as listed in Table 1. Experiments were conducted in the D1 low-speed wind tunnel at Beihang University. The free-stream velocity was fixed at $U_\infty = 20$ m/s, corresponding to the Reynolds number of $Re = 2.7 \times 10^5$ based on the root chord length. Aerodynamic forces were measured by a four-component strain-gage balance and the models were side-mounted on the balance. Based on an overall 95% confidence limits, the uncertainty for lift and drag components were less than 2% below $\alpha = 10°$, and around 1% beyond $\alpha = 10°$.

Fig. 15 Delta wing model
with LE tubercles (Chen
et al. 2013)

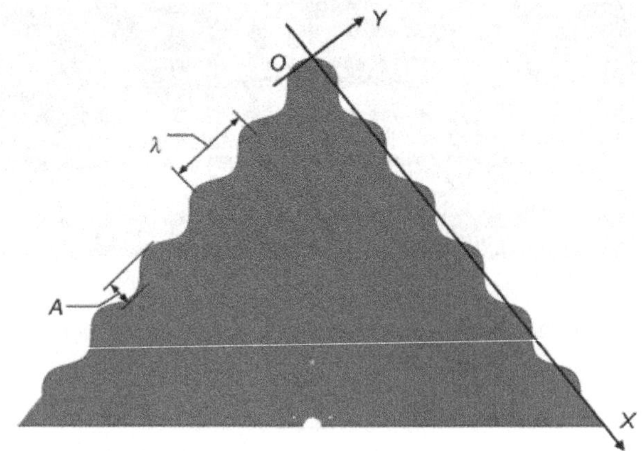

Table 1 The amplitude and
wavelength of different (LE)
configurations

Label	Amplitude (A) (% chord)	Wavelength (λ) (mm)
4S	2.5	63.5
4M	5.0	63.5
4L	12.0	63.5
6S	2.5	42.3
6M	5.0	42.3
6L	12.0	42.3
8S	2.5	31.7
8M	5.0	31.7
8L	12.0	31.7

Figure 16 shows the lift and drag coefficients of the delta wings with different amplitudes but the same wavelength of 42.3 mm. It can be observed that the lift coefficients of the tubercled wings are slightly lower than that of the smooth wing in the linear stage, and the maximum lift coefficient decreases with an increase in the tubercle amplitude. However, it is obvious that the stall angle is postponed by the LE tubercles, and the lift coefficient in the post-stall regime has also increased. Besides, the LE tubercles increase the drag coefficient of the delta wing for $\alpha > 26°$, which also increases with tubercle amplitude. Consequently, the lift-to-drag ratio of the tubercled wings becomes lower than that of the smooth wing. Larger amplitude may further decrease the lift-to-drag ratio.

Figure 17 shows the lift and drag coefficients of the delta wings with different wavelengths but the same amplitude of 5%c. It can be observed that the 4M case increases the maximum lift coefficient slightly by about 1%, while the other two cases of 6M and 8M decrease it by about 3% and 6%, respectively. When the wavelength is equal or larger than 42.3 mm, the stall angle is delayed, but all the three cases could increase the lift coefficient after stall in comparison with the smooth wing. The

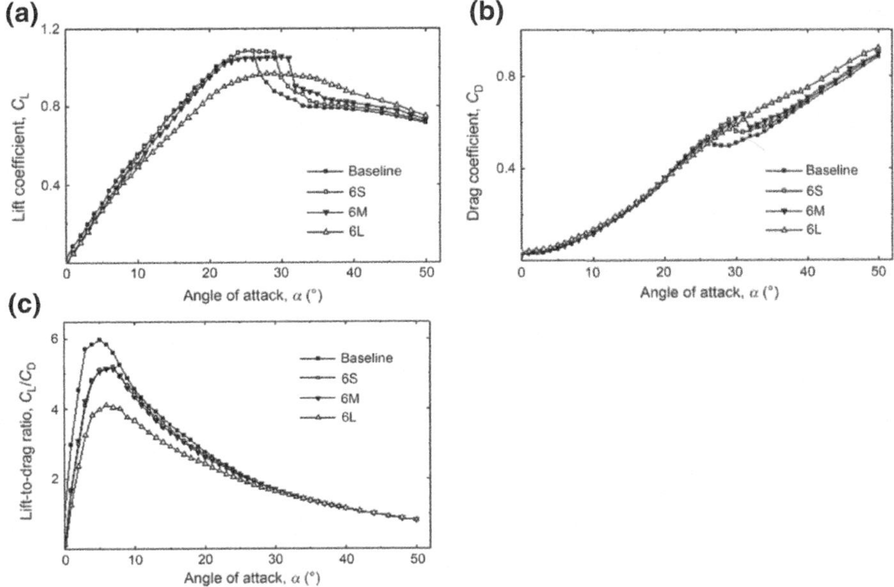

Fig. 16 Aerodynamic forces of the smooth and tubercled delta wing of "6" series (different amplitudes but the same wavelength of 42.3 mm). **a** Lift coefficient; **b** drag coefficient; **c** lift-to-drag ratio (Chen et al. 2013)

variations in the drag coefficient and the lift-to-drag ratio are also similar with that shown in Fig. 16. The LE tubercles may increase the drag coefficient while decreasing the lift-to-drag ratio in comparison with the smooth wing. Smaller wavelength leads to a larger deviation from the smooth wing.

Figure 18 shows the spatial development of the LE vortex for both smooth and 4M cases at $\alpha = 10°$, which are measured by stereoscopic particle-image velocimetry (SPIV). The left is the smooth case and the right is the 4M case. For the smooth wing, dual LE vortices can be clearly seen. For the 4M model, three LE vortices can be identified to form from each peak of the sinusoidal tubercles. The strength of the LE vortices for the 4M case is comparable to that for the smooth wing, which can help to explain the similar lift coefficient experienced for the smooth and tubercled wings at this AoA.

Figure 19 shows the surface oil pattern at $\alpha = 20°$ and 29° for the smooth and tubercled wings. At $\alpha = 20°$, three attachment lines (A1, A2, A3) and two separation lines (S2, S3) can be observed for the smooth wing, while only the primary attachment line and the secondary separation line are seen for the tubercled wing. The length of the primary attachment line for both the smooth and tubercled wings are comparable, indicating that the streamwise position of the primary vortex breakdown is almost the same for the two wings. Thus, the lift coefficient for the two wings is also nearly the same. At $\alpha = 29°$, only secondary separation line can be observed for the smooth wing, suggesting that the primary LE vortex has broke down. However, a pair of

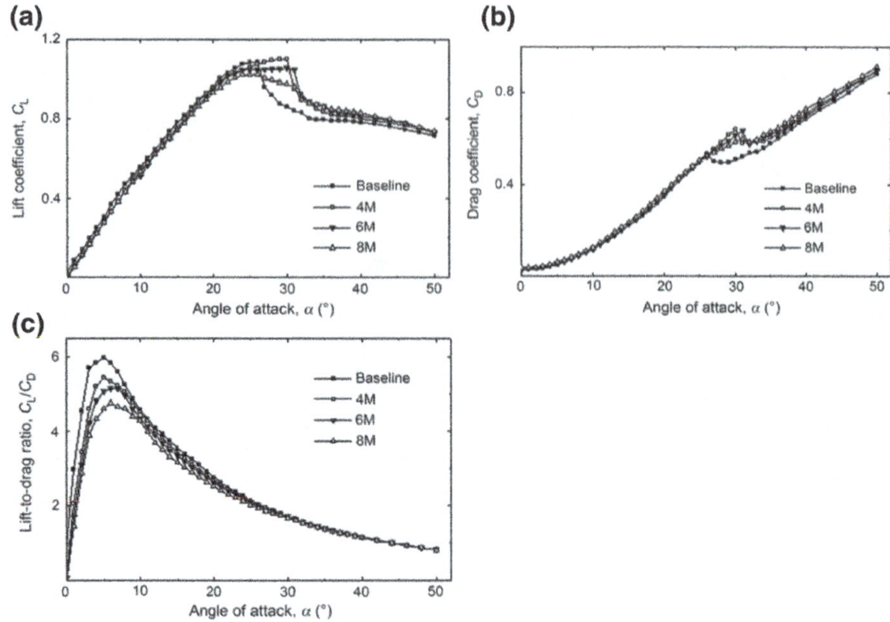

Fig. 17 Aerodynamic forces of the smooth and tubercled delta wing of "M" series (different wavelengths but the same amplitude of 5.0%*c*). **a** Lift coefficient; **b** drag coefficient; **c** lift-to-drag ratio (Chen et al. 2013)

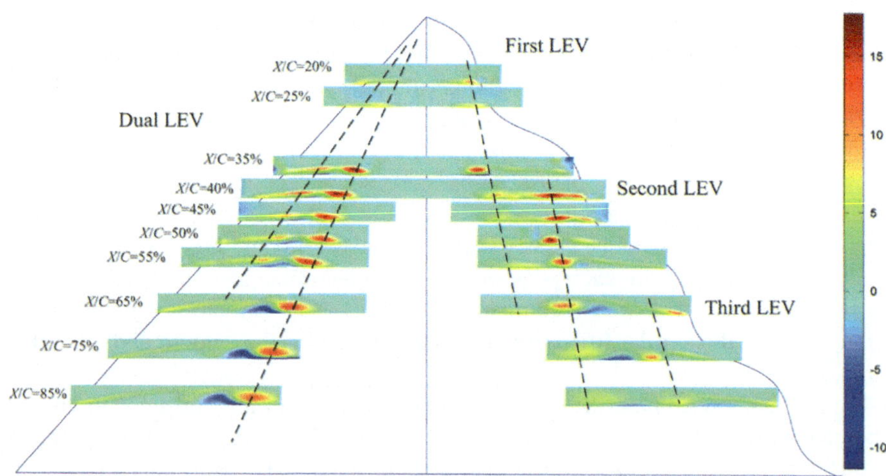

Fig. 18 Vortex structures of smooth (left part) and 4M tubercled (right part) delta wings at $\alpha = 10°$ (Chen and Wang 2014)

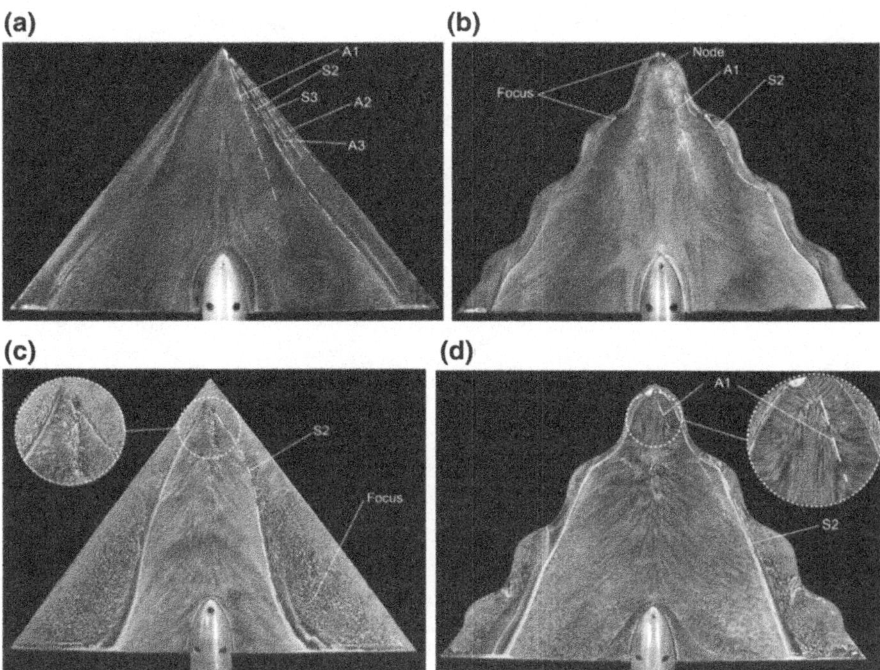

Fig. 19 Oil flow visualization for the smooth (**a, c**) and "4M" tubercled (**b, d**) delta wings at $\alpha = 20°$ (**a, b**) and $\alpha = 29°$ (**c, d**) (Chen et al. 2013)

short primary attachment lines can still be seen around the apex of the tubercled wing, suggesting that the complete breakdown of the primary vortex is delayed. From the surrounding area by the secondary separation lines, it is indicated that the separation region over the wing surface is reduced significantly by the LE tubercles. Thus, the tubercled wing exhibits a higher lift coefficient than the smooth wing.

Goruney and Rockwell (2009) conducted a PIV measurement of the near-surface flow field over a delta wing with LE tubercles. The peak-to-peak amplitude was denoted by φ and the wavelength was λ. The root chord length of the wing was $c = 200$ mm, the sweep angle was $50°$, and the wing thickness was 1.58 mm. The experiment was conducted at a free-stream velocity of $U_\infty = 70.9$ mm/s, corresponding to the Reynolds number of $Re = 1.5 \times 10^4$.

Figure 20 shows the time-averaged near-surface streamlines over the delta wing with different LE tubercles. For the smooth case, the focus of separation can be observed, indicating the existence of large-scale separation region over the delta wing surface. The focus of separation can also be found for the $\varphi/c = 0.01$, $\lambda/c = 0.05$ case. However, the focus of separation has changed into a focus of attachment or a node of attachment for other cases. As the amplitude and wavelength of the LE tubercles increase, the modification effects of the tubercles on the flow field are enhanced. For the case of $\varphi/c = 0.16$, $\lambda/c = 0.4$, in particular, we can find the occurrence of two foci of attachment, which form from each sinusoidal wave of the

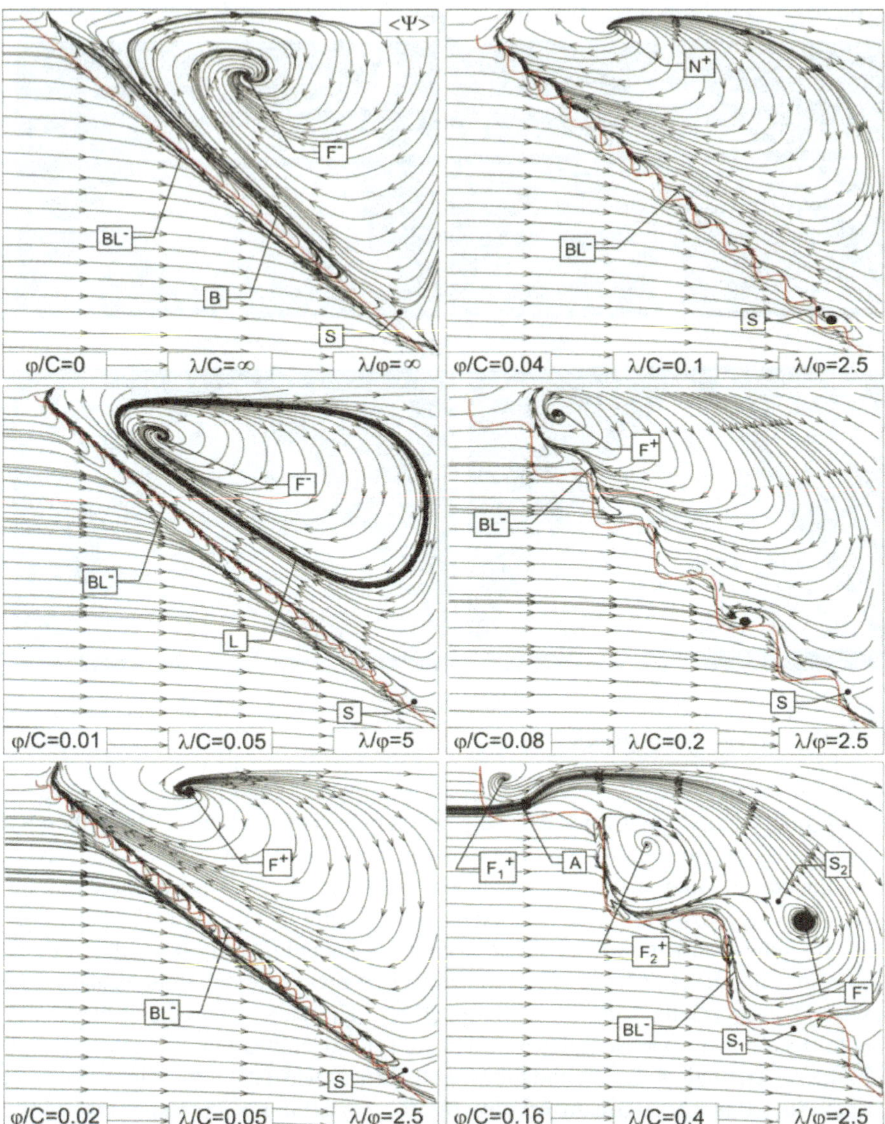

Fig. 20 Time-averaged streamlines on the near surface plane of delta wings with various values of amplitude φ and wavelength λ at $\alpha = 25°$ (Goruney and Rockwell 2009)

leading edge. Such variations in the flow topology mean significant changes to the flow field. One example is shown in Fig. 21 for time-averaged out-of-plane velocity on the near surface plane. It indicates that as the amplitude and wavelength of the LE tubercles increase, the vortical flow towards the wing suction surface is enhanced, which is beneficial for flow attachment.

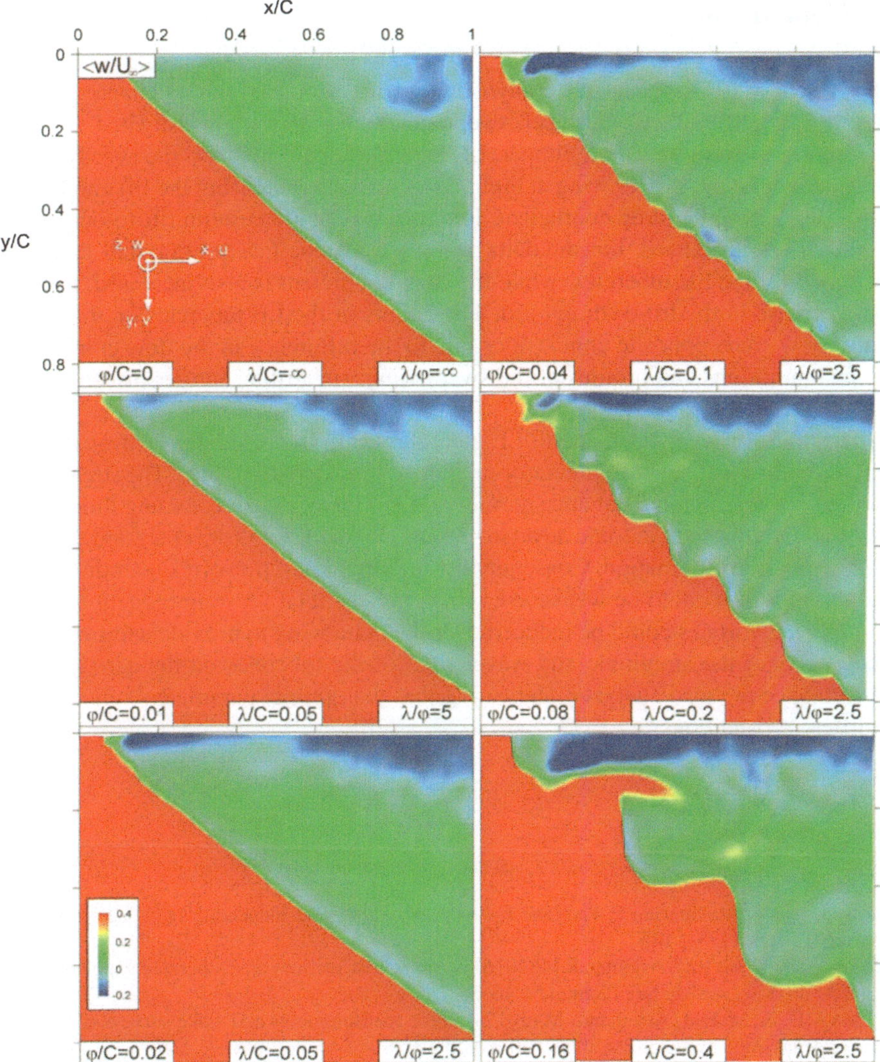

Fig. 21 Time-averaged out-of-plan velocity on the near surface plane of delta wings with various values of amplitude φ and wavelength λ at $\alpha = 25°$ (Goruney and Rockwell 2009)

5 Conclusion

The effects of LE tubercles on spanwise flow are introduced in this chapter, where their flow control effects on flipper, swept and delta wings are investigated.

Originated from the whale flipper, LE tubercles could improve the aerodynamic performance of the flipper wing at high AoAs, namely increasing the lift coefficient while decreasing the drag coefficient. Furthermore, the maximum lift-to-drag-ratio could also be increased. For the delta wing, the lift coefficient after stall could be increased by the LE tubercles, while the drag coefficient is also increased, leading to a decrease in the lift-to-drag ratio. The effects of the LE tubercles on the swept wing show some opposite conclusions due to the difference in the flow conditions, the configurations and profiles of the swept wing, suggesting further investigations are needed.

The flow control mechanism of LE tubercles is also revealed. It is proven that LE tubercles function as vortex generators to induce streamwise vortices. The streamwise vortices are beneficial to enhance momentum mixing of the surrounding fluids, thus enabling the flow to overcome adverse pressure gradient for flow separation delay or even separation elimination. Consequently, the strength of tip vortex over the flipper and swept wings could be weakened, which is beneficial for drag reduction, while multiple LE vortices could be induced over the delta wing to provide additional lift. The effects of the amplitude and wavelength of the tubercles on the aerodynamic performance are also analyzed, indicating that the optimal parameters might differ at different conditions.

References

Birch D, Lee T, Mokhtarian F, Kafyeke F (2004) Structure and induced drag of a tip vortex. J Aircraft 41(5):1138–1145

Bolzon MDP, Kelso RM, Arjomandi M (2016) Formation of vortices on a tubercled wing, and their effects on drag. Aerosp Sci Technol 56:46–55

Bolzon MD, Kelso RM, Arjomandi M (2017a) Force measurements and wake surveys of a swept tubercled wing. J Aerosp Eng 30(3):04016085

Bolzon MD, Kelso RM, Arjomandi M (2017b) Performance effects of a single tubercle terminating at a swept wing's tip. Exp Thermal Fluid Sci 85:52–68

Chen H, Pan C, Wang JJ (2013) Effects of sinusoidal leading edge on delta wing performance and mechanism. Sci China Technol Sci 56(3):772–779

Chen H, Wang JJ (2014) Vortex structures for flow over a delta wing with sinusoidal leading edge. Exp Fluids 55(6):1761

Custodio D, Henoch CW, Johari H (2015) Aerodynamic characteristics of finite span wings with leading-edge protuberances. AIAA J 53(7):1878–1893

Gursul I, Wang Z, Vardaki E (2007) Review of flow control mechanisms of leading-edge vortices. Prog Aerosp Sci 43(7):246–270

Goruney T, Rockwell D (2009) Flow past a delta wing with a sinusoidal leading edge: near-surface topology and flow structure. Exp Fluids 47(2):321–331

Mitchella AM, Délery J (2011) Research into vortex breakdown control. Prog Aerosp Sci 37(4):385–418

Miklosovic DS, Murray MM, Howle LE, Fish FE (2004) Leading-edge tubercles delay stall on humpback whale (*Megaptera novaeangliae*) flippers. Phys Fluids 16(5):39–42

Miklosovic DS, Murray MM, Howle LE (2007) Experimental evaluation of sinusoidal leading edges. J Aircraft 44(4):1404–1408

Pedro HTC, Kobayashi MH (2008) Numerical study of stall delay on humpback whale flippers. AIAA Paper 2008-0584

Spalart PR (1998) Airplane trailing vortices. Annu Rev Fluid Mech 30:107–138

van Nierop EA, Alben S, Brenner MP (2008) How bumps on whale flippers delay stall: an aerodynamic model. Phys Rev Lett 100(5):054502

Weber PW, Howle LE, Murray MM, Miklosovic DS (2011) Computational evaluation of the performance of lifting surfaces with leading-edge protuberances. J Aircraft 48(2):591–600

Wei Z, Lian L, Zhong Y (2018) Enhancing the hydrodynamic performance of a tapered swept-back wing through leading-edge tubercles. Exp Fluids 59(6):103

Effects of Leading-Edge Tubercles on Dynamically Pitching Airfoils

John Hrynuk and Douglas Bohl

Abstract The hydrodynamic effects of leading-edge (LE) tubercles, inspired by the flipper of humpback whales, has become a subject of interest for the control of flow over lifting surfaces. The primary focus of these efforts has centered on static cases where the angle-of-attack (AoA) of the flow is held constant. However, given the dynamic nature of the use of these flippers by humpback whales, there may also be benefit in dynamic scenarios as well. The investigation of flow control of dynamically actuated airfoils with LE tubercles is much more limited compared with the static studies. The results to date show that these bio-inspired airfoils expand the operating envelope for dynamic stall by enhancing lift and delaying stall to higher angles. These results indicate that applications such as on helicopter rotors and wind turbines may be able to avoid the detrimental effects of dynamic stall leading to more efficient devices, or to even better utilize its effects.

Keywords Dynamic stall · Pitching · Oscillations · Unsteady aerodynamics

1 Introduction

The field of aerodynamics can broadly be separated into "steady" and "unsteady" phenomena. In steady aerodynamics, a lifting surface is typically placed at a fixed angle-of-attack (AoA) and the aerodynamic properties (e.g. lift and drag) are measured. The AoA is then changed and held to allow the aerodynamic properties to be mapped out over a range of AoAs. Lift increases as the AoA increases until the static stall point and beyond which it decreases. This defines the static stall angle, α_{stall}. Drag also increases with increasing AoA and usually shows a significant increase after the stall angle. The static stall angle is typically between $\alpha_{stall} = 10°$ to $15°$ and

---

J. Hrynuk
US Army Research Lab, Aberdeen Proving Ground, MD 21005, USA

D. Bohl (✉)
Department of Mechanical and Aeronautical Engineering, Clarkson University, Potsdam, NY 13699-5725, USA
e-mail: dbohl@clarkson.edu

© Springer Nature Switzerland AG 2020
D. T. H. New and B. F. Ng (eds.), *Flow Control Through Bio-inspired Leading-Edge Tubercles*, https://doi.org/10.1007/978-3-030-23792-9_6

the maximum lift coefficient is nominally $C_{L,max} \approx 1$. Exact values depend on the shape of the lifting surface.

Lifting surfaces are not always held at a constant AoA, and the aerodynamic properties of lifting surfaces under dynamic motion can be quite different from static cases. For example, airfoils can be oscillated at small angles i.e. AoAs below the static stall angle. The rate of oscillation is characterized by the reduced frequency, $k = \pi c f / U_\infty$, where c is the chord length, f is the frequency of oscillation, and U_∞ is the free stream velocity. As the reduced frequency is increased, the wake profile changes and the airfoil begins to generate thrust instead of drag (Freymuth 1988; Triantafyllou et al. 1993; Dabiri 2005; Bohl and Koochesfahani 2009; King et al. 2018). These motions can also be characterized by the Strouhal number defined as $St = f a_{te} / U_\infty$. Here, a_{te} is the length of the sweep of the trailing-edge (TE) of the airfoil. The Strouhal number is sometimes used as it accounts for the amplitude of motion, which is lacking in the reduced frequency. Regardless, this pitching motion is a simplified model of how fish swims (King et al. 2018; Schouveiler et al. 2005; Anderson et al. 1998).

A second "classic" dynamic condition is to either pitch or oscillate the airfoil well past the static stall angle. This is commonly known as dynamic stall. Dynamic stall is a fundamental flow event that occurs when a wing undergoes a rapid change in AoA past α_{stall}. It is commonly characterized using a dimensionless pitch rate, $\Omega^* = \dot{\alpha} c / U_\infty$ where $\dot{\alpha}$ is the rate of change in wing AoA (rad/s), c is the wing chord length, and U_∞ is the incoming freestream velocity for constant pitch rate motions. Dynamic stall can also be observed during large amplitude oscillating conditions, which are also characterized by the Strouhal number and/or reduced frequency as previously defined. During a dynamic stall event, the flow remains attached well past the static stall point and a LE vortex or dynamic stall vortex (DSV) is generated at a high enough AoA. While the flow is technically separated from the airfoil surface (i.e. it does not follow the surface of the airfoil), the presence of the DSV lowers the pressure on the suction side of the wing, generating a significant increase in lift. The DSV eventually convects away from the airfoil surface and the airfoil stalls upon losing the enhanced lift. Typically, dynamic stall angles and maximum lift coefficients are 3–4 times larger than the static conditions (Gendrich 1999).

Dynamic stall has been most commonly associated with helicopter rotor blades (Carr et al. 1977; McCroskey 1981), however more recent studies have evaluated dynamic stall on wind turbine blades, which may be able to use these dynamics for increased power production (Leishman 2002; Larsen et al. 2007; Yen and Ahmed 2013; Pierce and Hansen 1995; Fujisawa and Shibuya 2001; Schreck and Robinson 2007; Holierhoek et al. 2013). Helicopter rotors experience dynamic stall because of the blade rotation, which causes a cyclic change in the relative AoA due to the changes in the contribution of the forward flight speed and rotational speed of the rotor to the apparent velocity vector. Control of dynamic stall is a focus of many studies (Barwey and Gaonkar 1994; De Gregorio 2012; Gardner et al. 2013; Biava et al. 2012; Conlisk 2001; Yu et al. 1995) as it is a major factor in limiting forward flight speed of rotor craft. Dynamic stall also impacts turbomachinery due to rotor-stator interaction and was one of the original driving forces to understanding dynamic stall (Ekaterinaris and Platzer 1997; Carr et al. 1977).

Historically, static aerodynamics has been studied first and in more depth. This was due to the use of airfoils for wings on planes, as control surfaces for ships, etc., where they are more likely to encounter fixed AoAs by design. The study of dynamic properties of lifting surfaces began as a result of rotor powered aircraft and turbo machinery where the lifting surfaces encounter variable AoAs, also by design. This has more recently expanded to other technical applications like wind turbines, UAV's, unmanned submersible vehicles, etc. as their use has increased. The control of dynamic stall is critical in these applications as it typically limits their performances (Carr et al. 1977; Ekaterinaris and Platzer 1997; Geissler et al. 2004, 2005; Mai et al. 2008).

The study of the effects of bio-inspired tubercled airfoils based on the flippers of humpback whales has followed a similar path in that the static properties of tubercled wings have been studied first and more extensively. The review of these studies is the subject of other chapters in this book and not covered in detail here. When considering the dynamics and application of these bio-inspired airfoils, it is important to realize that humpback whales are known to be highly dynamic swimmers (Fish and Battle 1995; Fish 1999; Segre et al. 2017). The flippers are used as control surfaces for swimming and observation of the flippers while swimming shows that they are used in dynamic, rather than static, conditions. While interesting flow behaviors have been observed for static tubercled wings, study of dynamic motion of these wings is also insightful. To date, the study of the effects of dynamic actuation of bioinspired wings with LE tubercles has been limited.

There is good reason to believe that beneficial effects from dynamic stall may be present for humpback whales. Segre et al. (2017) studied humpback whales in the wild and noted some interesting aspects of humpback whale behavior: "Immediately before opening their mouths, humpbacks will often rapidly move their flippers, and it has been hypothesized that this movement is used to corral prey." They further suggested that the large flippers and dynamic motions might be used to generate significant lunging forces. Segre et al. (2017) applied suction cup digital recording tags to several humpback whales and were able to record two instances of what they called "hydrodynamically active flipper-strokes." One such motion is of particular interest to the evaluation of flippers as dynamic, rather than static, lift generators. While the dynamic behavior of humpbacks is obvious upon watching videos of the whales swimming, Segre et al. (2017) documents a clear case which can be evaluated aerodynamically. They recorded a rapid downward flipper stroke that actuated the flippers 90° in approximately 0.8 s, as shown in Fig. 1. Given the conditions noted here and a few reasonable assumptions, this motion is easily shown to be in the range of the unsteady aerodynamic conditions encompassing dynamic stall.

The actuation motion observed in Segre et al. (2017), combined with the mean chord length of the humpback whale from Fish and Battle (1995) of 0.82 m, and an assumption of 3 m/s forward velocity of the whale, leads to a non-dimensional pitch rate of $\Omega^* = 0.21$. This result places the motion significantly above the threshold

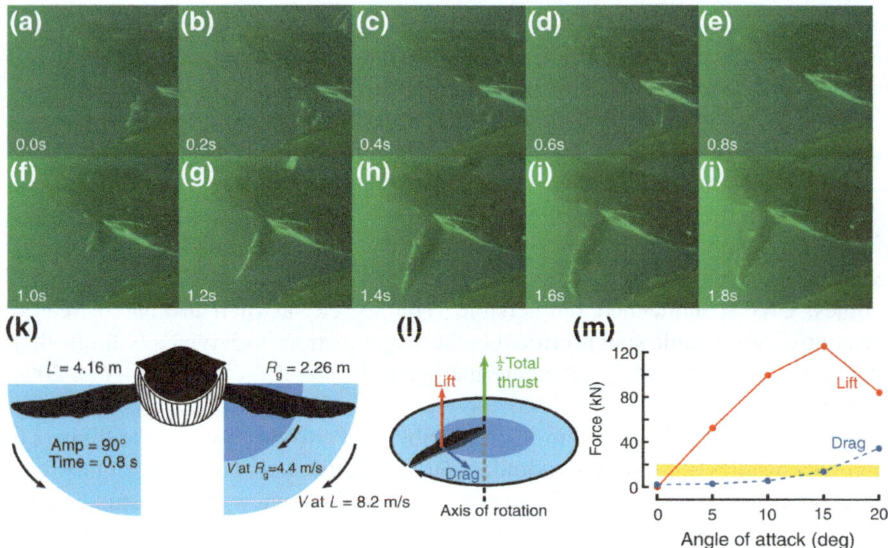

Fig. 1 Hydrodynamically active maneuver of the humpback whale from Segre et al. (2017)

of $\Omega^* > 0.05$ with a peak AoA greater than the static stall angle, which defines the presence of a dynamic stall event. Segre et al. (2017) assumed a static lift maximum of 120 kN for this particular humpback. Results of the dynamic stall study by Gendrich (1999) showed that the unsteady force on an infinite span wing was three times the maximum static lift. If instead of using the $C_{l,max}$ of the static case, one conservatively assumes a more realistic $2 * C_{l,max}$ for the dynamic case, the whale could be generating an unsteady peak lunging force roughly equivalent to its own body weight. With this "back of the envelope" result in mind, the evaluation of the effect of tubercles on dynamic stall becomes even more interesting.

2 Review of Tubercle Effects on Static Wings

To fully understand the effects of tubercled wings undergoing dynamic stall, the flow field effects of tubercles on static wings must first be briefly discussed. Stanway (2008) studied tubercles as a possible factor in improving propulsive efficiency of flapping foils and turning efficiency using the wing model from Miklosovic et al. (2004). Likewise, in other studies, particle-image velocimetry (PIV) data was used to calculate the surface normal vorticity, as shown in Fig. 2, and quantitatively showed the vortex structures (Custodio 2007; Miklosovic et al. 2004).

Stanway (2008) suggested that tubercles act similarly to the LE on delta wings, where flow wraps around the LE to generate vortical structures. Like delta wings, airfoils with tubercles generate counter-rotating vortex regions due to the non-uniform

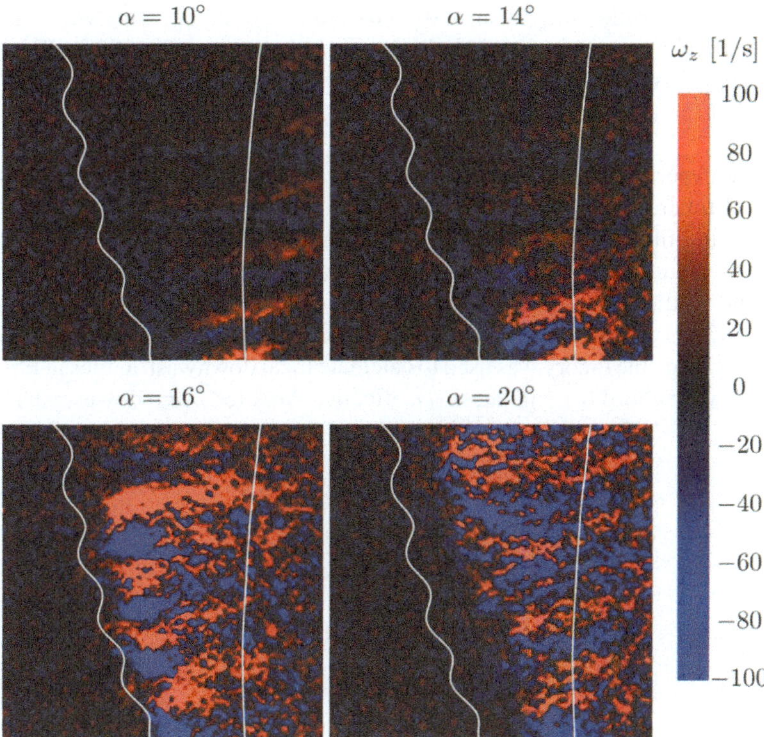

Fig. 2 Surface normal vorticity. Modified from Stanway (2008)

LE. In some regions, these vortices drive the flow downward toward the surface, reattaching the flow at high incidence while causing flow separation in other regions. This effect results in a delay in stall and a resulting net increase in lift at high AoA compared to a standard airfoil for the static case. Stanway (2008) also explored thrusting behaviors by heaving and pitching their wing at a wide range of Strouhal numbers and maximum pitching angles. The Strouhal number used in Stanway was defined as $\text{St} = fh_c/U_\infty$, where h_c is a normalized heave amplitude based on the mean chord as the wing was not rectangular. The wings were also pitched with peak angles from $10°$ to $60°$ over the Strouhal number range of $\text{St} = 0.2$ to 0.8, which is a known thrust generating region (Bohl and Koochesfahani 2009; King et al. 2018). Note that the wing in Stanway (2008) was pitched with a non-linear motion profile so the traditional metrics of reduced frequency (k) or pitch rate (Ω^*) were not used.

Thrust performance in Stanway (2008) was found to be less effective for the tubercled wing when compared to a baseline wing. However, the numerical study by Wang et al. (2014) observed no benefit or detriment to thrusting using a tubercled wing compared to a baseline case. One key difference between the study in Stanway (2008) and Wang et al. (2014) is that the mode used by Stanway (2008) was a finite span, fin-shaped wing while Wang et al. (2014) used a tubercled LE wing with repeating boundary conditions (i.e. an "infinite" span airfoil). Stanway's (2008) PIV results on static wings quantitatively showing the flow field, including the vortex structures, showed that the theories of prior works on the existence of vortex structures were correct.

Despite experimental observations of vortex generation by Stanway (2008) and theories of tubercles as vortex generators (Fish and Battle 1995; Miklosovic et al. 2004), the simplified model of van Nierop (2009) and van Nierop et al. (2008) showed many of the same lift and drag characteristics as experimental results. This model applied the Kutta-Joukowski lift theorem, where lift was calculated as $L = \rho U_\infty \Gamma$, to tubercled LE wings. Circulation (Γ) was calculated using the Kutta condition in Eq. 1, and lifting-line theory was used to calculate local downwash angles in Eqs. 2–3, which use wing chord (c), thickness (t), effective AoA (α^e), and downwash (w).

$$\Gamma = -\pi U_\infty \left(c + \frac{4}{3\sqrt{3}} t \right) \alpha^e + \mathcal{O}\left(\left(\alpha^e \right)^3, t^3 \right) \tag{1}$$

$$\alpha^e = \alpha - \frac{w}{U_\infty} \tag{2}$$

$$w = \frac{1}{4\pi} \int\limits_{s}^{-s} \frac{\mathrm{d}\Gamma/\mathrm{d}y'}{y' - y} \mathrm{d}y' \tag{3}$$

The integrals in this study (Van Nierop 2009; Van Nierop et al. 2008) were solved both numerically and with a series expansion, which could define the tubercled LE. Close similarities between this analytical model and prior experiments suggest one of two possible conclusions: (1) Flow modification by tubercles can be mostly explained by LE shape alone, or (2) one or more of the model assumptions, notably the inherent spanwise flow limitation, reflected the flow behavior being imposed by the LE shape. An alternative way to view the first cause was presented in Van Nierop (2009) and van Nierop et al. (2008), who suggested that the pressure distribution was modified in such a way that boundary layer separation behind the peak of the tubercles was delayed, thus delaying stall. The pressure field modification argument is consistent with the development of the vertical structures due to the shape of the LE. Specifically, it can be shown that the presence of a pressure gradient is required to transport vorticity at a surface, where it is generated, into a flow field (Lighthill 1963; Bohl and Foss 1999). Numerical results (Watts and Fish 2001; Stanway 2008) suggested that tubercles inhibited spanwise flow over a wing. An experimental study by Cai et al. (2018) using a single LE tubercle near the mid-span of a finite span wing was able to isolate the spanwise flow limitation originally posited by Watts and Fish (2001). They were able to show that separation occurring on one half of the wing did not transfer to the still attached flow opposite the tubercle on the LE.

The cause of flow modifications due to LE tubercles on static wings/airfoils can be summarized as one or a combination of several competing theories. Specifically, tubercles have been thought to act in the following ways:

(1) As vortex generators adding momentum to the flow and reattaching the boundary layer (Fish and Battle 1995; Miklosovic et al. 2004)
(2) As vortex lift generators similar to delta wings (Fish and Battle 1995; Miklosovic et al. 2004; Stanway 2008; Johari et al. 2007)
(3) As a boundary layer separation delay mechanism (Van Nierop et al. 2008).
(4) As spanwise flow limiting devices (Watts and Fish 2001; Stanway 2008; Cai et al. 2018)

3 Fundamental Dynamic Stall with Tubercled Wings

Many of the theorized benefits of tubercles from the static studies have been utilized in the past as passive control mechanisms for dynamic stall. Some applying research using vortex generators, which have similar behaviors as tubercled wings, was performed by Geissler et al. (2004, 2005), Mai et al. (2008) and Heine et al. (2011). Geissler et al. (2004, 2008) conducted experiments with a morphing LE while Mai et al. (2008) and Heine et al. (2011) applied small vortex generating disks to the underside of rotor blades undergoing dynamic stall. While Geissler's work is less reflective of the tubercles, modifying the LE by drooping it downward did have an effect on dynamic stall. Mai and Heine's vortex generators, which had a very positive effect towards suppressing the dynamic stall vortex on a helicopter rotor blade, might act in a similar fashion to the tubercles in Stanway (2008). It should be noted however that vortex generators used to control dynamic stall are typically much smaller in scale (i.e. on order of a boundary layer thickness) than the LE tubercles, which are scaled on the order of the chord length. Passive flow control strategies typically exhibit a performance cost in flow regimes outside of those for which they are needed. Much of the static force coefficient data for tubercled airfoils shows little or no performance penalty below the stall point (Miklosovic et al. 2004), which makes them an interesting candidate for actual applications.

Despite the thorough study of tubercles as to their effects on static wings, the work by Segre et al. (2017) shows that static consideration of tubercles alone likely does not lead to a full understanding of their benefits. Although the work on dynamic wings with tubercles is still limited, it has so far been found to have rich dynamics that merit much further study.

Borg (2012) experimentally investigated the performance effects of tubercles on dynamic stall, specifically looking at oscillatory pitching motion relevant to helicopter blades. A sinusoidal oscillating motion profile was constructed around two mean angles, 3° below and 5° above the static stall point of $\alpha_{stall} = 16°$, with pitching amplitudes of 5° and 7°. All the pitching motions in Borg (2012) had a reduced frequency of $k = 0.08$, but the small pitching amplitudes kept the flow in a "light" dynamic stall regime. Light dynamic stall refers to a dynamic stall where the static stall angle is only exceeded by a few degrees, whereas deep dynamic stall studies typically exceed the static stall angle by more than double. The results in Borg (2012) displayed similar trends to earlier works (Carr et al. 1977) with smaller amplitude motions creating lower increases in dynamic lift (beyond the static case) and larger amplitude motions producing larger increases in dynamic lift. Performance losses due to hysteresis were also increased with increased pitching amplitude beyond the static stall point. This behavior is the expected result in that once a dynamic stall vortex is generated, it will convect away from the wing and ultimately cause a more pronounced hysteresis loop.

In the light dynamic stall cases, where the wings were pitched only a few degrees beyond the static stall angle, the lowest amplitude tubercles produced more dynamic lift and experienced lower hysteresis losses compared to the baseline wing. Figure 3 contains a comparison of two figures from Borg (2012), showing the baseline wing hysteresis loops and the large amplitude (0.12c) mid-wavelength (0.33c) tubercles. High amplitude tubercles increased peak lift by 42–44% beyond the static case, but had very large hysteresis losses, on the order of 60%. In comparison, Borg (2012) reported the baseline wing oscillated at 5° exceeded static lift by 33% but lost upward of 80% of the static peak lift during the hysteresis loop. Note that the percent changes were calculated with respect to the individual case peak lift and not the overall best-case value. Large spanwise wavelength tubercles produced the most lift but had the highest hysteresis losses. Conversely, low wavelength tubercles produced lower peak lift but had lower hysteresis losses. For their higher amplitude dynamic stall experiments, where the wings were oscillated to angles 5° above the static stall point, Borg suggested that tubercled wings had no improvement over a baseline case. While this likely refers to the lift being generated at those angles, the hysteresis behavior of tubercled wings had significant benefits compared to straight LE wings.

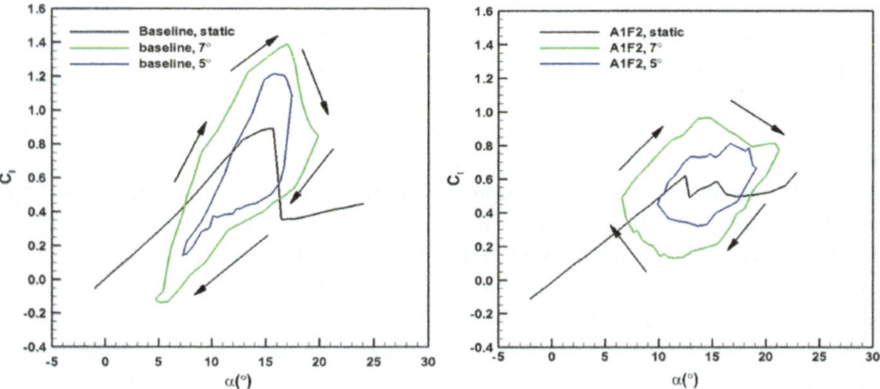

Fig. 3 Lift coefficient as a function of AoA for the baseline and bio-inspired/tubercled wing. Data used from Borg (2012)

Further analysis of the results from Borg (2012) in Fig. 3 shows that the trends of the static tubercled wing, which had a lower peak lift but a softer static stall behavior compared to the baseline wing, also extended to the light dynamic stall cases. For both dynamic cases, the peak lift was reduced but the lift drop-off and the hysteresis loop was softer for the tubercled wing compared to the baseline case. Note that the arrows in Fig. 3 show the direction of dynamic lift during the pitching motion.

The most thorough study to date on the fundamental flow fields associated with tubercled wing dynamic stall was done by Hrynuk (2015). In these experiments, the flow fields on the suction side of a tubercled airfoil with 4% amplitude tubercles was compared to a baseline straight LE wing at $Re_c = 12,000$ using Molecular Tagging Velocimetry (MTV), a molecular equivalent to PIV. Experiments were performed with a NACA0012 cross-section to allow comparison to previous dynamic stall data. In these experiments, the airfoil was pitched from $0°$ to $55°$ with a constant pitch rate motion with non-dimensional pitch rates of $\Omega^* = 0.1, 0.2$ and 0.4. Four measurement planes, two aligned with the peak and trough and two equally spaced between (i.e. "near peak" and "near trough"), were investigated.

Analysis of the baseline wing data in Fig. 4 showed the typical features of the dynamic stall process originally described in Carr et al. (1977). First, a region of separation forms at the TE of the airfoil and moves forward towards the LE while the flow remains attached at the LE even though the airfoil is past the static stall angle (Fig. 4, $15°$). The boundary layer eventually separates from the LE and forms a local peak in the vorticity, which develops into the DSV (Fig. 4, $25°$). Again, it is interesting to note that the flow reattaches downstream of the LE separation point. Near the TE, the flow is clearly reversed along the airfoil surface at this angle. However, the flow continues to follow the airfoil surface above the separation point and remain "attached" to the airfoil at this point. The flow field is then dominated by the DSV, which is assumed to be the primary dynamic lift generating mechanism (Fig. 4, >$25°$), though a smaller shear layer vortex is also present. The interaction

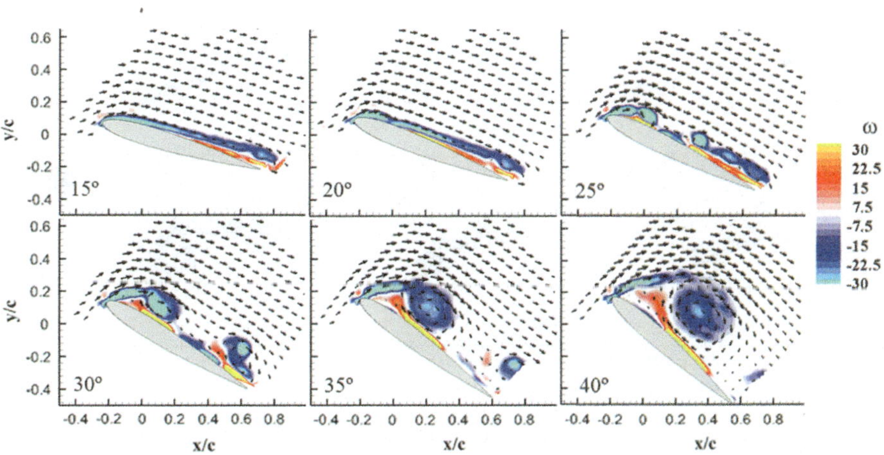

Fig. 4 Baseline wing dynamic stall evolution (Hrynuk 2015)

of these flow structures, as well as the reverse flow boundary layer below the DSV, was shown to control the convective behavior of the DSV (Gendrich 1999; Hrynuk 2015). As the pitching motion progresses, the DSV continues to grow in size while convecting downstream. While the flow over the airfoil is clearly separated (as defined based on the flow field) along the entire chord length, the presence of the DSV near the surface provides the low pressure that allows the lift force to remain high. Eventually, the DSV moves downstream from the wing at the same time a near-LE secondary recirculating vortex lifts off the surface. After the pitching motion completes, a shedding mode exists until the wing is pitched back below the static stall angle.

When tubercles were applied to the wing, several changes in the flow fields were observed by Hrynuk (2015). The flow conditions observed for static tubercled wings (Watts and Fish 2001), where flow was accelerated through the troughs and three-dimensional effects changed the lift behavior, carried over to the dynamic case. Figures 5 and 6 show the flow fields of the dynamic stall event for a peak and trough plane of a tubercle. Over the peak of a tubercle, the DSV formed further back along the chord and at a higher angle than the baseline wing, essentially delaying dynamic stall over this region. These observations are consistent with the "downwash" caused by the streamwise vorticies formed by the tubercles. In contrast, the trough region of the tubercled wing formed a DSV earlier than the baseline and formed closer to the LE compared to the baseline. These observations are consistent with the "upwash" caused by the streamwise vorticies formed by the tubercles. The shape of the DSV and its location along the chord impacted the strength and formation of the secondary boundary layer and shear layer vortices associated with the dynamic stall process. The lifting of the secondary boundary layer under the DSV was absent for the tubercled wing at all span locations. The lifting of this secondary boundary layer has been shown to be a key trigger of the separation of the DSV from the proximity of the

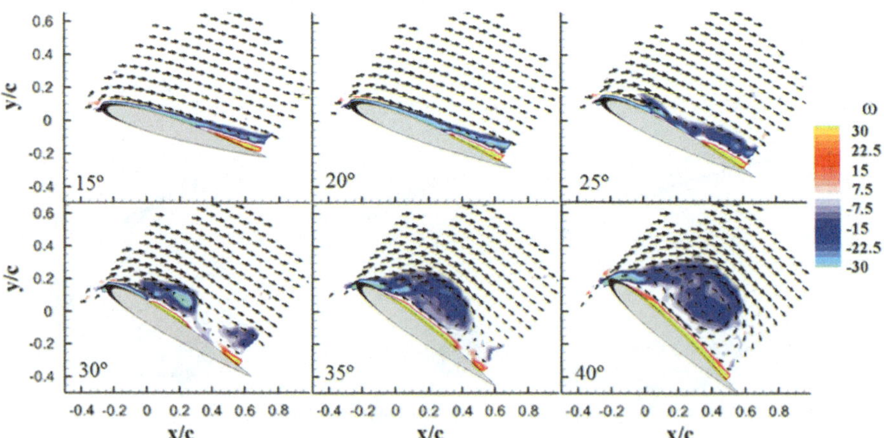

Fig. 5 Tubercled wing peak plane dynamic stall evolution (Hrynuk 2015). Grey shows airfoil shape in the trough, black indicates shape of the baseline airfoil for comparison

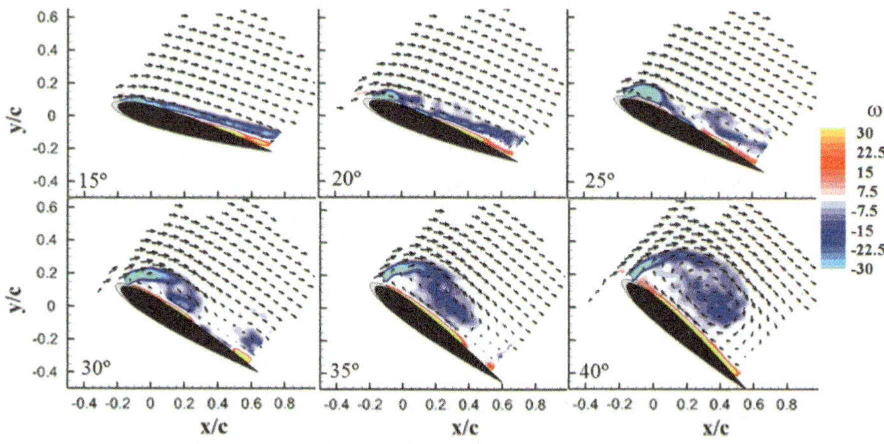

Fig. 6 Tubercled wing trough plane dynamic stall evolution (Hrynuk 2015). Grey shows airfoil shape in the trough, black indicates shape of the baseline airfoil for comparison

airfoil surface (Gendrich 1999) and therefore appears to be the reason for the delay in the convection of the DSV.

The location and strength of the dynamic stall vortex are known to be the primary factors affecting dynamic lift on a wing (Gendrich 1999). Tracking of the vortex in Hrynuk (2015) showed that DSV formation occurred at lower angles and further forward for the trough plane and at higher angles and further back over the peak when compared to the baseline case. It is interesting to note that the DSV in the trough plane moved downstream to align with the peak plane DSV shortly after forming. The DSV for the tubercled airfoil remained closer to the suction surface than the baseline throughout the entire pitching motion, including later times as it convected downstream.

The convection speed of the DSV was calculated from the tracking data and showed the DSV accelerating and decelerating differently for each plane and the baseline through this convection process. The convective speed of the vortex for the baseline straight wing was shown by Hrynuk (2015) to be related to the interaction with the shear layer vortices. The boundary layer behavior and formation of shear layer vortices was also shown to be a function of the dimensionless pitch rate. Changes in this behavior caused the dynamic stall vortex convection speed to remain nearly constant as a function of pitch rate. Convective velocity calculation showed that the dynamic stall vortex for the tubercled wing convected away from the wing slower than the baseline on average for low pitch rates, but higher for higher pitch rates.

Hrynuk (2015) did not measure the lift and drag directly. However, the effects of the tubercles on lift were inferred using the circulation, Γ, of the DSV which can be related to lift through the Kutta-Joukowski lift theorem. This theorem essentially demonstrates that higher circulation correlates to a higher lift. Results showed that the tubercled wing consistently had a higher circulation than the baseline, regardless of pitch rate. The highest benefit from the tubercles was at the lowest pitch rate tested, where the circulation was 20% higher than the baseline wing.

It was concluded that the combination of the closer and stronger dynamic stall vortex generated by the tubercled wing likely generated higher dynamic lift forces and maintained them to higher angles. As a result, there might be significant benefits to systems where increased dynamic lift would be highly beneficial.

4 Applied Performance Benefits of Wings Experiencing Dynamic Stall

A report by Howle (2009) on the Whale Power wind turbine blade provided some support to the theory presented in Hrynuk (2015). This report showed that a small horizontal axis wind turbine outfitted with LE tubercles produced significantly more power over the course of several months. Two competing theories existed for the increase in power: (1) the static lift behavior of tubercles was providing more force when the baseline blade would have been stalled, or (2) rapid changes in wind direction were causing dynamic stall events which produced the higher power output.

Horizontal axis wind turbines are historically the most common wind turbines, typically with three blades rotating in a propeller like fashion. Conversely, vertical axis wind turbines use vertical fixed blades rotating around a central axis, see Fig. 7. In 2017, Wang and Zhuang performed a computational study on the performance of a vertical axis wind turbine which may help settle the competing theories on improved wind turbine performance with tubercles. In this study, Wang and Zhuang (2017) applied LE tubercles to the spinning blades of a vertical turbine, which was always known to be effected by dynamic stall in a similar fashion to helicopter rotor blades.

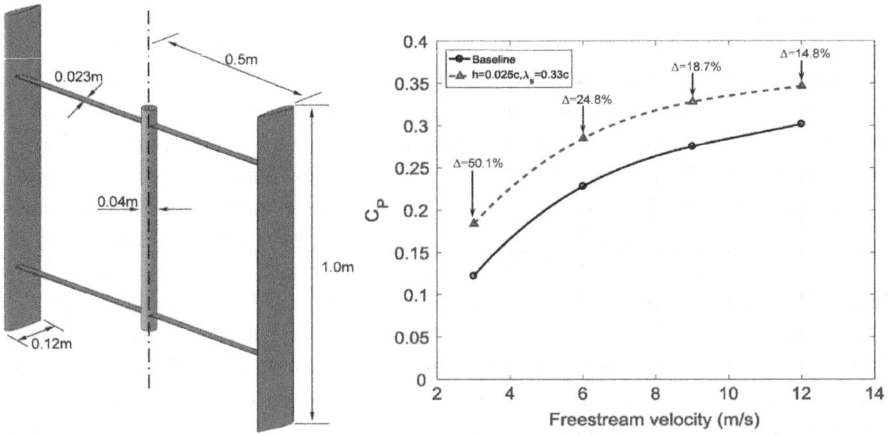

Fig. 7 Vertical axis wind turbine model and power curve comparison from Wang and Zhuang (2017)

Wang and Zhuang (2017) showed a power increase at all free stream velocity cases for their tubercled turbine in Fig. 7. As dynamic stall is known to be present in the vertical axis turbines, this suggests that the results of Hrynuk (2015) are likely the better explanation of the results in Howle (2009), which showed that power increases for a horizontal axis wind turbine.

Several of the findings in their computational study were also very similar to those observed in Hrynuk (2015). For instance, Wang and Zhuang (2017) state: "It can be clearly seen from (Fig. 18b, Wang and Zhuang 2017), slight flow separation can be found on the trough- and middle-plane, but no separation is observed on the peak-plane." Figure 8 is Fig. 18b from Wang and Zhuang (2017) and shows the differing vortex formation behaviors over baseline and tubercled wings. This result mirrors the results found in Hrynuk (2015), with early trough separation but attached flow near the peak region. Wang and Zhuang (2017) go on to conclude that the streamwise vortex pair generated by the tubercle peak allows for reverse flow only behind the trough, while that same vortex structure suppressed flow separation near the peak. Ozen and Rockwell (2010) observed similar counter-rotating pairs for a rapidly flapping wing with a tubercled LE but did not study if any LE vortex was present. These results seemingly extend the findings in Watts and Fish (2001) as well as many others (Miklosovic et al. 2004; Stanway 2008; Johari et al. 2007; Cai et al. 2018) to the dynamic problem, where they affect the dynamic stall vortex formation.

Contrary to these findings, a study by Bai et al. (2005) observed decreased performance for a tubercled vertical axis wind turbine. However, it should be noted that Bai et al. (2005) used a 2D and 2.5D CFD methods. Studies like the one by

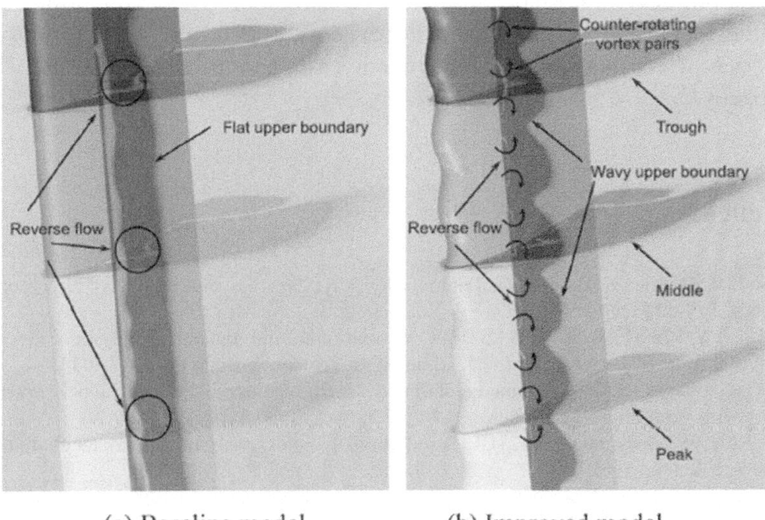

(a) Baseline model   (b) Improved model

Fig. 8 Recreated Fig. 18b from Wang and Zhuang (2017) which shows the instantaneous vorticity distributions at three planes along the span of each wing and one spanwise plane located at approximately ¼ chord

Tabib et al. (2017) on horizontal wind turbines showed that in regions with large three-dimensional effects, the 2D and 2.5D models break down, which may be the cause of poor performance observed in Bai et al. (2005).

5 Conclusion

While the study of tubercled LE wings experiencing dynamic stall events is limited, the results so far suggest that there may be significant flow control benefits for this condition. Results in Hrynuk (2015) and Wang and Zhuang (2017) showed the flow field effects and each suggest more dynamic lift is generated when tubercles are applied to the LE of wings due to changes in the DSV formation and dynamics. Engineering design results, such as those in Howle (2009), showed some of the benefits from tubercled wings in unsteady flows. However, unlike the static wing case which has large quantities of data for a variety of different tubercle shapes and sizes, dynamic pitching studies have looked at relatively few cases.

The understood physics of tubercles from static studies, like those of Watts and Fish (2001) and Miklosovic et al. (2004), appear to extend in some ways to dynamic studies. The early vortex formation in troughs observed in Hrynuk (2015) was strikingly similar to the early separation from Watts and Fish (2001), while vortex pairs theorized in Miklosovic et al. (2004) and observed in Stanway (2008) were observed in the dynamic case by Wang and Zhuang (2017). However, study of dynamic motion of lifting surfaces for standard airfoils and wings has shown that one cannot always extend the physics of the static case to the dynamic one. These similarities for the bioinspired airfoils/wings suggest that the dynamic effects of tubercled wings are at least as rich, if not more so, for the dynamic case than the static case and warrant further study.

References

Anderson JM, Streitlien K, Barrett DS, Triantafyllou MS (1998) Oscillating foils of high propulsive efficiency. J Fluid Mech 360:41–72

Bai CJ, Lin YY, Lin SY, Wang WC (2005) Computational fluid dynamics analysis of the vertical axis wind turbine blade with tubercle leading edge. J Renew Sustain Ener 7(3):033124

Barwey D, Gaonkar GH (1994) Dynamic-stall and structural-modeling effects on helicopter blade stability with experimental correlation. AIAA J 32(4):811–819

Biava M, Khier W, Vigevano L (2012) CFD prediction of air flow past a full helicopter configuration. Aerosp Sci Technol 19:3–18

Bohl DG, Foss JF (1999) Near exit plane effects caused by primary-plus-secondary tabs. AIAA J 37(2):192–201

Bohl DG, Koochesfahani MM (2009) MTV measurements of the vortical field in the wake of an airfoil oscillating at high reduced frequency. J Fluid Mech 620:63–88

Borg J (2012) The effect of leading edge serrations on dynamic stall. PhD Thesis University of Southampton

Cai C, Zuo Z, Morimoto M, Maeda T, Kamada Y, Liu S (2018) Two-step stall characteristic of an airfoil with a single leading-edge protuberance. AIAA J 56(1):64–77

Carr LW, McAlister KW, McCroskey WJ (1977) Analysis of the development of dynamic stall based on oscillating airfoil experiments. NASA Technical Note D-8382

Conlisk AT (2001) Modern helicopter rotor aerodynamics. Prog Aero Sp Sci 37:419–476

Custodio D (2007) The effect of humpback whale-like leading edge protuberances on hydrofoil performance. Masters Thesis Worcester Polytechnic Institute

Dabiri JO (2005) On the estimation of swimming and flying forces from wake measurements. J Exp Biol 208(18):3519–3532

De Gregorio F (2012) Flow field characterization and interactional aerodynamics analysis of a complete helicopter. Aero Sp Sci Tech 19:19–36

Ekaterinaris JA, Platzer MF (1997) Computational prediction of airfoil dynamic stall. Prog Aerosp Sci 33:759–846

Fish FE (1999) Performance constraints on the maneuverability of flexible and rigid biological systems. In: Proceedings of the eleventh international symposium of unmanned untethered submersible technology, pp 394–406

Fish FE, Battle JM (1995) Hydrodynamic design of the Humpback Whale flipper. J Morphol 225:51–60

Freymuth P (1988) Propulsive vortical signature of plunging and pitching airfoils. AIAA J 26:881–883

Fujisawa N, Shibuya S (2001) Observations of dynamic stall on Darrieus wind turbine blades. J Wind Eng Ind Aerod 89:201–214

Gardner AD, Richter K, Mai H, Altmikus ARM, Klein A, Rohardt CH (2013) Experimental investigation of dynamic stall performance for the EDI-M109 airfoils. J Am Helicopter Soc 58(1):1–13

Geissler W, Dietz G, Mai H, Junker M (2004) Dynamic stall control investigations on a full size chord blade section. 29th European Rotorcraft Forum, Friedrichshafen Germany

Geissler W, Dietz G, Mai H (2005) Dynamic stall on a supercritical airfoil. Aerosp Sci Technol 9:390–399

Gendrich CP (1999) Dynamic stall of rapidly pitching airfoils: MTV experiments and Navier-Stokes simulations. PhD Dissertation Michigan State University

Heine B, Mulleners K, Joubert G, Raffel M (2011) Dynamic stall control by passive disturbance generators. AIAA J 51(9):2086–2097

Holierhoek JG, De Vaal JB, Van Zuijlen AH, Bijl H (2013) Comparing different dynamic stall models. Wind Energy 16:139–158

Howle LE (2009) Report on the efficiency of a Whalepower Corp. 5 meter prototype wind turbine blade, Bellequant LLC, Durham NC

Hrynuk JT (2015) The effect of leading edge tubercles on dynamic stall. Ph.D. Dissertation Clarkson University

Johari H, Henoch C, Custodio D, Levshin A (2007) Effects of leading-edge protuberances on airfoil performance. AIAA J 45(11):2634–2642

King JT, Kumar R, Green MA (2018) Experimental observations of the three-dimensional wake structures and dynamics generated by a rigid, bioinspired panel. Phys Rev Fluids 3:034701

Larsen JW, Nielsen SRK, Krenk S (2007) Dynamic stall model for wind turbine airfoils. J Fluids Struct 23:959–982

Leishman JG (2002) Challenges in modeling the unsteady aerodynamics of wind turbines. In: 21st ASME wind energy symposium and the 40th AIAA Aerospace Sciences meeting, Reno NV

Lighthill MJ (1963) Introduction to boundary layer theory. In: Rosenhead L (ed) Laminar boundary layers. Clarendon, Oxford

Mai H, Dietz G, Geissler W, Kichter K, Bosbach J, Richard H, de Groot K (2008) Dynamic stall control by leading edge vortex generators. J Am Helicopter Soc 53(1):26–36

McCroskey WJ (1981) The phenomenon of dynamic stall. NASA Technical Memorandum 81264

Miklosovic DS, Murray MM, Howle LE, Fish FE (2004) Leading-edge tubercles delay stall on humpback whale (*Megaptera novaeangliae*) flippers. Phys Fluids 16(5):39–42

Ozen A, Rockewell D (2010) Control of voritcal structures on a flapping wing via a sinusoidal leading-edge. Phys Fluids 22:021701

Pierce K, Hansen AC (1995) Prediction of wind turbine rotor loads using the Beddoes-Leishman model for dynamic stall. J Sol Energy Eng 117:200–204

Schouveiler L, Hover FS, Triantafyllou MS (2005) Performance of flapping foil propulsion. J Fluid Struct 20:949–959

Schreck SJ, Robinson MC (2007) Horizontal axis wind turbine blade aerodynamics in experiments and modeling. IEEE T Energy Conver 22(1):61–70

Segre PS, Mdudzi S, Meyer MA, Findlay KP, Goldbogen JA (2017) A hydrodynamically active flipper-stroke in humpback whales. Curr Biol 27:R623–R641

Stanway MJ (2008) Hydrodynamic effects of leading-edge tubercles on control surfaces and in flapping foil propulsion. Masters Thesis MIT

Tabib M, Rasheed A, Siddiqui MS, Kvamsdal T (2017) A full-scale 3D Vs 2.5D Vs 2D analysis of flow pattern and forces for an industrial-scale 5 MW NREL reference wind-turbine. Energy Procedia 137:477–486

Triantafyllou GS, Triantafyllou MS, Grosenbaugh MA (1993) Optimal thrust development in oscillating foils with application to fish propulsion. J Fluid Struct 7:205

Van Nierop EA (2009) Flows in films and over flippers. Ph.D. thesis Harvard University

Van Nierop EA, Alben S, Brenner MP (2008) How bumps on whale flippers delay stall: an aerodynamic model. Phys Rev Lett 100(5):054502

Wang Y, Hu W, Zhang S (2014) Performance of the bio-inspired leading edge protuberances on a static wing and a pitching wing. J Hydrodyn 26(6):912–920

Wang Z, Zhuang M (2017) Leading-edge serrations for performance improvement on a vertical-axis wind turbine at low tip-speed-ratios. Appl Energy 208:1184–1197

Watts P, Fish FE (2001) The influences of passive leading edge tubercles on wing performance. In: 12th International Symposium on Unmanned Untethered Submersible Technology, Durham NH

Yen J, Ahmed NA (2013) Enhancing vertical axis wind turbine by dynamic stall control using synthetic jets. J Wind Eng Ind Aerod 114:12–17

Yu YH, Lee S, McAlister W, Tung C, Wang CM (1995) Dynamic stall control for advanced rotorcraft application. AIAA J 33(2):289–295

Effects of Leading-Edge Tubercles on Structural Dynamics and Aeroelasticity

Bing Feng Ng, Edwin Jit Guan Ong, Rafael Palacios and T. H. New

Abstract The design of tubercles has both aerodynamics and structural considerations. In this chapter, we discuss structural design, stability and aeroelasticity of lifting surfaces that are modified with leading-edge (LE) tubercles. With LE tubercles, bending and torsional frequencies are slightly lower due to the combined effects of spanwise stiffness and inertia variations from spanwise chord length variation and mass redistribution. In the pre-stall regime, lifting surfaces are likely to encounter instability (flutter) due to flexibility and rear placement of the centre-of-gravity. The structural dynamics and unsteady aerodynamics with LE tubercles have opposite influence on margins-of-stability with the latter having the dominant effect. Numerical investigations show that the flutter speed is consistently mildly higher with LE tubercles and they have reduced effect on the margins for stability when concentrated inboard of the wing or on sweptback wings as the sweep angle is increased.

Keywords Structural dynamics · Bending · Torsion · Flutter · Aeroelasticity · Stability

1 Introduction

From the previous chapters, it is realized that leading-edge (LE) tubercles are able to delay stall and improve post-stall aerodynamics, with further potential for acoustic reduction (Van Nierop et al. 2008; Watts and Fish 2001; Zhang and Frendi 2016). Motivated by these optimistic findings, there has been significant interest to adopt LE tubercles on industrial blades as well as aircrafts wings (Corsini et al. 2013; Fish 2009; Goruney and Rockwell 2009; Howle 2009; Malipeddi et al. 2012) to benefit from improved aerodynamic performances in the post-stall regime (Miklosovic

B. F. Ng (✉) · E. J. G. Ong · T. H. New
School of Mechanical and Aerospace Engineering, Nanyang Technological University, 50
Nanyang Avenue, 639798 Singapore, Singapore
e-mail: bingfeng@ntu.edu.sg

R. Palacios
Department of Aeronautics, Imperial College London, South Kensington Campus, London SW7
2AZ, UK

et al. 2004). However, majority of the blades/wings are slender and require further investigations into their stability margins as a result of LE tubercles.

Most studies have reported that LE tubercles are able to delay stall through the generation of vortical structures that promote spanwise flow and momentum exchange within the turbulent boundary layer (Miklosovic et al. 2004). On fixed wings, these vortical structures create non-uniform downwash and upwash on the tubercle crests and troughs, respectively (Wei et al. 2015). On flapping wings, the presence of LE tubercles delivers lower thrust and efficiency as the vortical structures interfere with thrust wake from flapping (Ozen and Rockwell 2010).

In the design of tubercles, most studies have considered sinusoidal functions of uniform amplitude and wavelength for the LE (Hansen et al. 2011). In general, the amplitude of tubercles has more influence on post-stall performances over wavelength. Tubercles with larger amplitude exhibit progressive stalling, lower pre-stall and maximum lift while having higher post-stall lift (Hansen 2012). Tubercle amplitude and wavelength can also be optimized to obtain configurations that take advantage of improved post-stall performances without adversely affecting pre-stall lift (Hansen et al. 2012). Another design parameter for tubercles is the point along a tubercle where it is initiated and terminated on the wing (phase of tubercles) which may considerably alter lift and induced drag (Bolzon et al. 2016).

Even though LE tubercles are shown to be beneficial in terms of the aerodynamics, an in-depth investigation is also required from the aeroelastic perspective as both structures and aerodynamics are affected by the tubercles in different ways (Ng et al. 2016a, 2017a). Here, we discuss the structural dynamics and aeroelastic stability margins of generic lifting surfaces modified with LE tubercles. We will first evaluate geometrical considerations in the design of LE tubercles, followed by an in-depth analysis on the effects of tubercles on natural modes of vibration. Subsequently, stability will be discussed by means of an aeroelastic tool (Murua et al. 2012) to determine flutter speeds. Different tubercles distribution and wing swept angles will also be examined.

2 Structural Considerations of LE Tubercles

In the structural design of lifting surfaces to accommodate LE tubercles, the amplitude, wavelength and phase of tubercles play critical roles in affecting the natural modes of vibration. Another important factor is the cross-section of the lifting surface in which the tubercles are to be implemented. On most lifting surfaces, the cross-section is of airfoil configuration where mass is concentrated ahead of the mid-chord. Consequently, modifications to the LE to accommodate tubercles will result in significant chordwise and spanwise material removal or addition that re-distributes mass and re-aligns centre-of-gravity (C.G.) positions, affecting both structural inertias and stiffnesses. This is clearly illustrated in Fig. 1 where the design of LE tubercles on the lifting surface with airfoil cross-section results in material addition and removal in the peak and trough regions, respectively.

Fig. 1 Design of LE tubercles on lifting surfaces with airfoil cross-section. Illustrating the material addition and removal in the areas of peak and trough, respectively

The phase of the tubercles is critical in determining the resulting performance of the lifting surface. The phase of tubercles refers to the point along the tubercle where the wing terminates at the root or tip. Aerodynamically, it has been shown that the phase of the tubercles have the largest effects on lift coefficient, induced drag coefficient and lift-to-drag ratio (Bolzon et al. 2016). Structurally, the existence of a peak at the root or tip would result in root stiffening or increased tip mass, respectively. These have counteracting effects on increasing or reducing stability margins.

As shown in Fig. 2, starting the phase of tubercles with zero phase at root (reference to a sine function) would result in reduced tip mass while strengthening the root. The opposite can be said for starting with a π phase lead at the root. On the other hand, starting with a $\pi/2$ or $3\pi/2$ phase lead at the root would either increase or decrease, respectively, both the local chord at the root and tip.

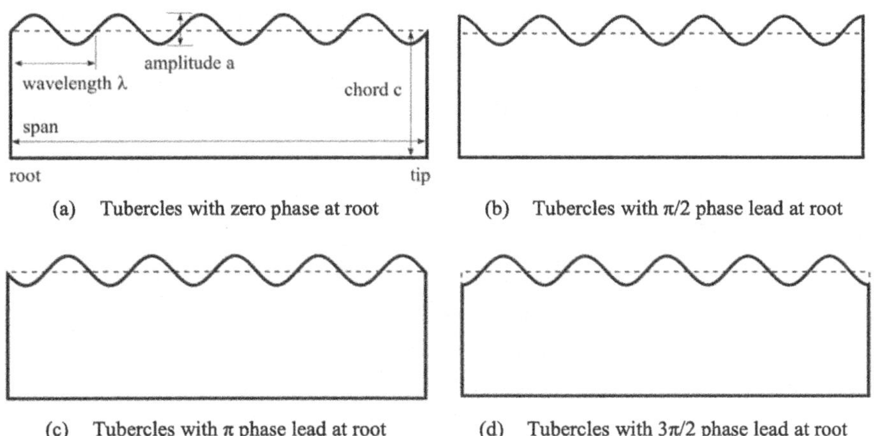

Fig. 2 Spanwise distribution of LE tubercles with different starting phase

Fig. 3 Illustration of lifting surfaces of airfoil cross-section without tubercles and with tubercles, including that of the flat plate with tubercles

In general, airfoils differ markedly by their location of maximum thickness, thickness to chord ratio, camber, etc. and the design of tubercles along the LE would result in significantly different material distribution. From an aeroelastic point of view in evaluating structural responses through different tubercle configurations, it is however more representative to consider rectangular cross-sections (flat plates) as shown in Fig. 3.

Here, the Goland wing (Goland 1945) that is a widely-used aeroelastic benchmark configuration, is adopted as the baseline lifting surface. The wing has a span of 6.10 m, chord measuring 1.83 m and its structure is modelled as a flexible beam. The properties of the Goland wing are shown in Table 1. The first bending and torsion frequencies are 7.66 and 15.2 Hz, respectively, while the flutter speed at sea-level straight flight has been reported to be around 166 m/s (Ng et al. 2015; Wang et al. 2006).

To understand the extent of flat plates as a generic model in determining the response of lifting surfaces with LE tubercles, two models, one with S833 airfoil cross-section and a second one with rectangular cross-section are compared for their natural frequencies of vibration with the same planform area. The results are computed using ANSYS Mechanical (ANSYS® Academic Research Mechanical, Release 18.1, n.d.). Figure 4 shows the normalized first bending frequency (with respect to a baseline without tubercles) for two and ten LE tubercles of zero and π phase lead configurations. While the difference in normalized bending frequencies between the airfoil and rectangular cross-sections amplify as tubercle amplitude

Table 1 Properties of the Goland wing

Chord (m)	1.83
Half-span (m)	6.10
Flexural axis from LE (normalized by chord)	0.33
Centre of gravity from LE (normalized by chord)	0.43
Moment of inertia (kg m)	8.64
Mass per unit length (kg/m)	35.7
Torsional stiffness (N m^2)	0.99×10^6
Bending stiffness (N m^2)	9.77×10^6

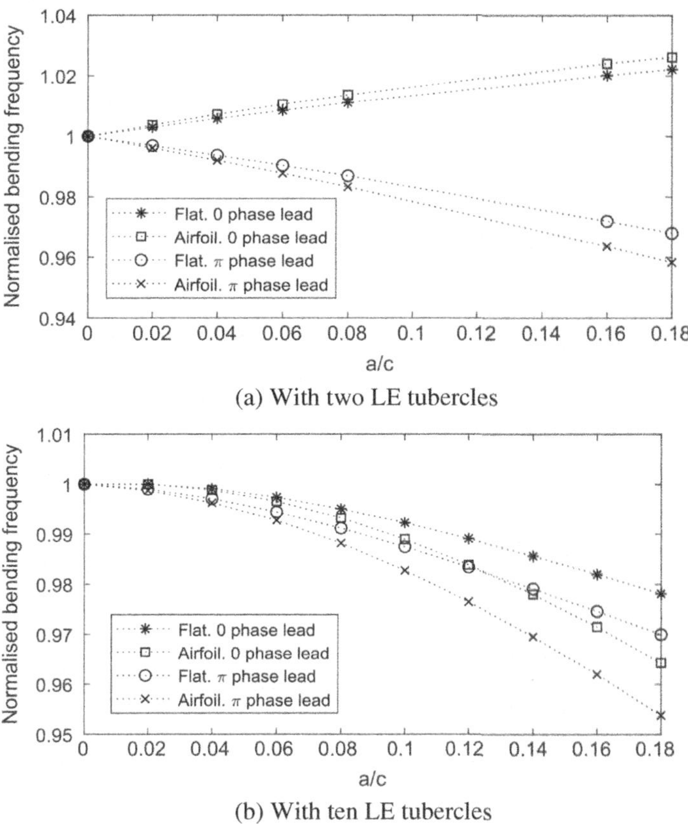

(a) With two LE tubercles

(b) With ten LE tubercles

Fig. 4 Comparison of normalised first bending frequency for Goland wing with airfoil and rectangular cross-sections for zero and π phase lead configurations

increases, the trends are largely the same. In general, we observe decreasing bending frequencies as tubercle amplitude increase, which is the result of a multitude of factors including added tip mass, root stiffening, C.G. movement, etc. Detailed analysis will be provided in subsequent sections.

In the case of normalized torsional frequencies in Fig. 5, the two cross-sections also exhibit similar trends with increasing amplitude. It is to be emphasized that the comparison made is between a lifting surface with S833 airfoil and a flat plate, and the S883 airfoil is considerably thick fore of the mid-chord where tubercles are positioned. Should other airfoil configurations be used, it is anticipated that the qualitative trends would be similar but the quantitative differences will be dependent on the cross-sections. As a case study in this chapter, the flat plate will be considered as representative in providing estimations of frequency responses in the presence of

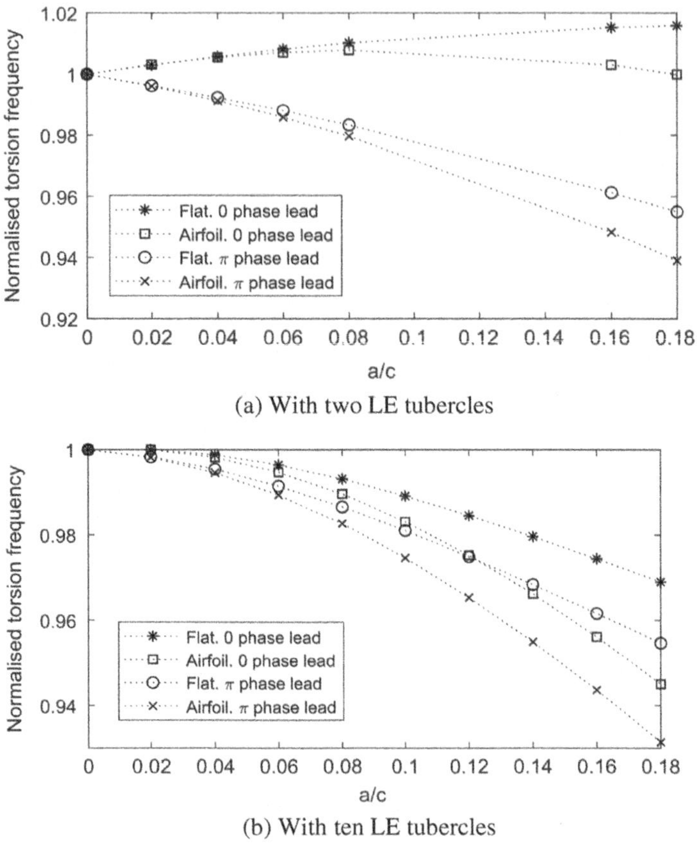

(a) With two LE tubercles

(b) With ten LE tubercles

Fig. 5 Comparison of normalised first torsion frequency for Goland wing with airfoil and rectangular cross-sections for zero and π phase lead configurations

LE tubercles, neglecting the added dimension of thickness that is dependent on the airfoil selection.

3 Effects of Tubercles on Structural Modes of Vibration

The effects of tubercle amplitude, wavelength and phase on the natural modes of vibration will be discussed in this section using the flat plate model of the Goland wing. Despite having the same number and amplitude of tubercles, structural responses will differ according to the phase of tubercles. This is the result of C.G. movement and also the distribution of peaks and troughs along the span that affects local chord lengths. We will first analyse the default zero and π phase leads.

3.1 Tubercles with Zero Phase Lead

For zero phase lead, the root begins with the tubercle midway between a peak and trough, which is immediately followed by a tubercle peak, as illustrated in Fig. 2a. Close to the tip, we have a tubercle trough followed by termination midway between the trough and peak. Displacement plots for the first, second bending and torsion modes for two tubercles are shown in Fig. 6, computed using ANSYS Mechanical (ANSYS® Academic Research Mechanical, Release 18.1, n.d.). Despite increasing tubercle amplitudes from a/c = 0.02 to a/c = 0.16, the displacement plots are largely similar. For the first and second bending with small tubercle amplitude, the maximum displacement occurs mid-chord at the tip of the wing, which is subsequently shifted to the trailing-edge (TE) as tubercle amplitude increases. For the first torsion, the maximum displacement occurs at the LE and TE at the tip of the wing, which is shifted to the LE at the tip as tubercle amplitude increases.

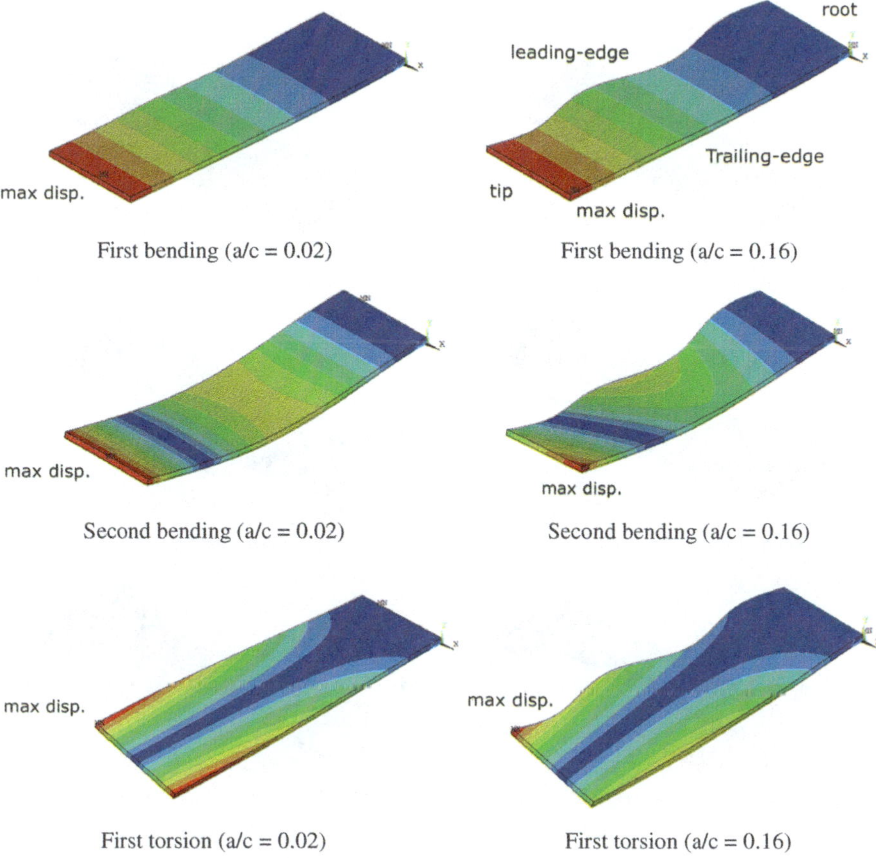

Fig. 6 Displacement plots for the first, second bending and torsion modes for two tubercles

For six LE tubercles, the displacement plots for the first, second bending and torsion modes are shown in Fig. 7. For the first and second bending with small tubercle amplitude, the maximum displacement occurs close to mid-chord at the tip of the wing, which are respectively, shifted towards and away from the LE as tubercle amplitude increases. For the first torsion, the maximum displacement occurs at the LE and TE at the tip of the wing. As tubercle amplitude increases, the maximum is shifted to the tubercle peak closest to the tip instead of occurring at the tip itself. At the tip and with zero phase lead, the tubercles terminate in between a peak and trough.

The effects of increasing tubercle amplitude from a/c = 0 to a/c = 0.18 on the first and second bendings are illustrated in Figs. 8 and 9, respectively. The five plots are for different number of tubercles ranging from two (λ/c = 1.67) to ten tubercles (λ/c = 0.33) in increments of two tubercles.

For zero phase lead, as tubercle amplitude increases, bending frequencies exhibit an initial increase, which subsequently falls as amplitude is increased further. This is the same across all tubercle wavelengths. The inverted U-shaped trend can be

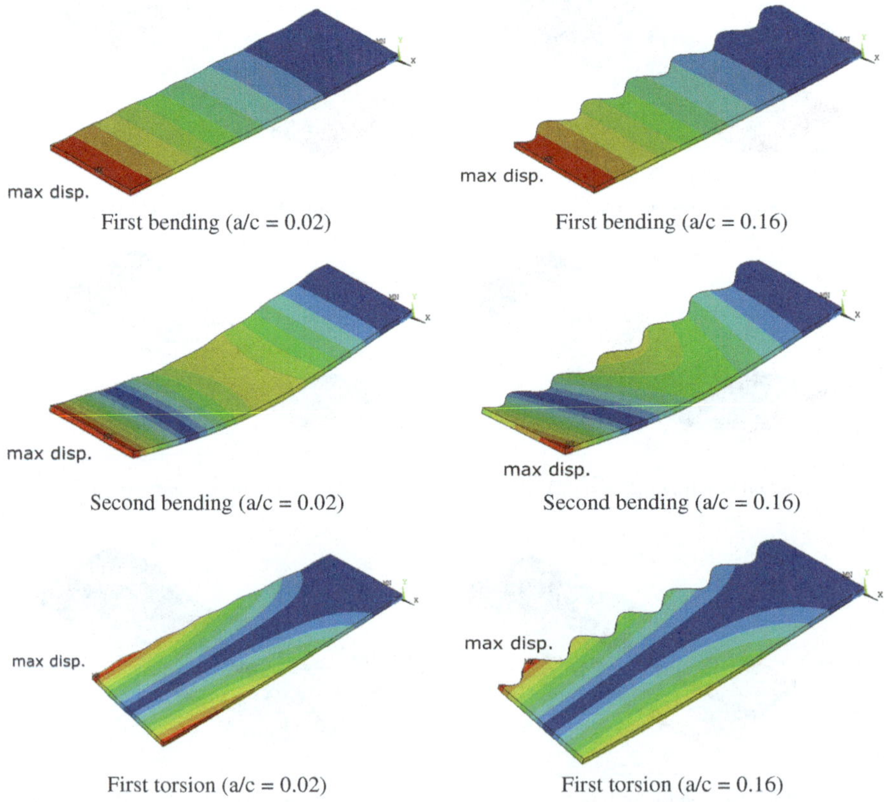

First bending (a/c = 0.02)   First bending (a/c = 0.16)

Second bending (a/c = 0.02)   Second bending (a/c = 0.16)

First torsion (a/c = 0.02)   First torsion (a/c = 0.16)

Fig. 7 Displacement plots for the first, second bending and torsion modes for six tubercles

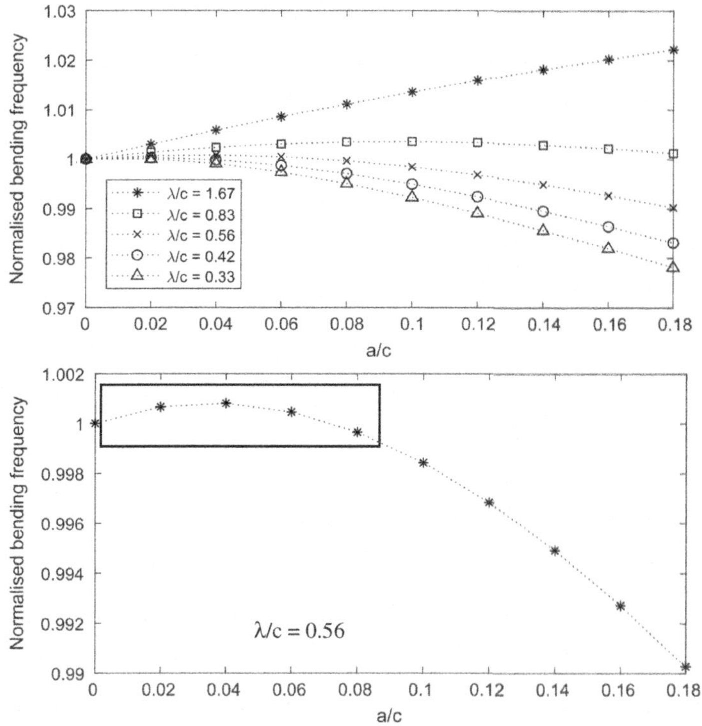

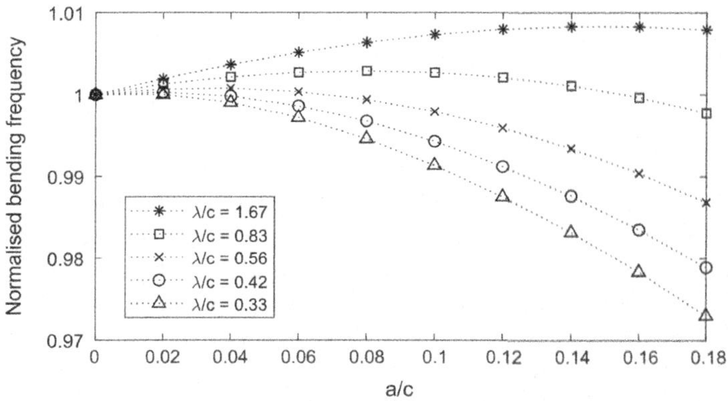

Fig. 8 Effect of increasing tubercle amplitude on first bending frequency (zero phase lead). Bottom plot is for wavelength of $\lambda/c = 0.56$ (6 tubercles)

Fig. 9 Effect of increasing tubercle amplitude on second bending frequency (zero phase lead)

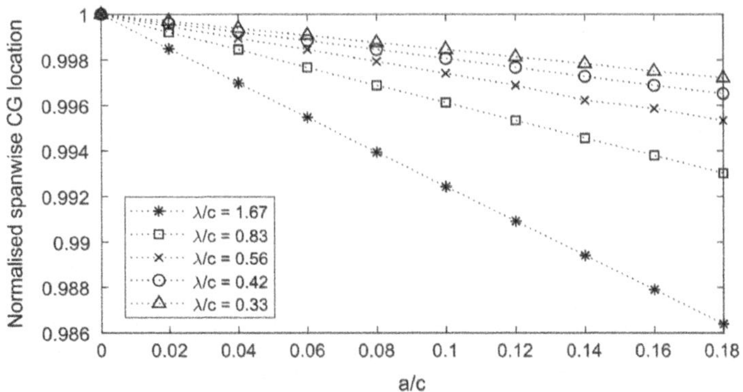

Fig. 10 Normalised spanwise C.G. location for increasing tubercle amplitude. Normalised with respect to mid-span (zero phase lead)

observed more clearly when plotted individually for $\lambda/c = 0.56$. This initial increase in frequency depends on the number of tubercles and are less pronounced with more tubercles (or wavelength reduced).

There are few reasons that may result in the trends observed. For zero phase lead, bending frequencies exhibit an initial increase with amplitude as the presence of the trough at the tip reduces chord length and hence planform area. This increases bending frequencies. In addition, the peak towards the root region leads to stiffening that has a positive effect on increasing bending frequencies. The C.G. location also plays a significant role in increasing bending frequencies. As the tubercle amplitude increases, the spanwise location of the C.G. moves towards the root, as shown in Fig. 10.

However, as the amplitude of tubercles increases further, the effects that lead to increased frequencies are overcome by reduced chord lengths at trough regions along the span of the blade. This causes the blade to become more flexible to reduce bending frequencies that result in the overall inverted U-shape trend.

The first torsional frequency also exhibits the same initial increase followed by a drop as we increase the number of tubercles. The first torsional frequency is shown in Fig. 11. The initial increase can be seen as the effects of root stiffening from the tubercle peaks but is later offset by reduced chords at the trough regions.

For increasing number of tubercles (reduction in wavelength), we observe a relatively linear decrease in frequencies for both bending and torsional modes from Figs. 12. With two tubercles ($\lambda/c = 1.67$), larger amplitude leads to higher initial frequencies due to root stiffening. However, as the number of tubercles is increased, the factors resulting in the initial higher frequencies are offset by an increasingly even spanwise distribution of tubercles. The rate of decrease is also more significant for tubercles of larger amplitude, resulting in the intersections between the plots of different amplitudes. The decrease in frequencies can be understood from the spanwise C.G. positioning, which moves toward the tip of the blade as number of tubercles is increased (wavelength reduced) in Fig. 13.

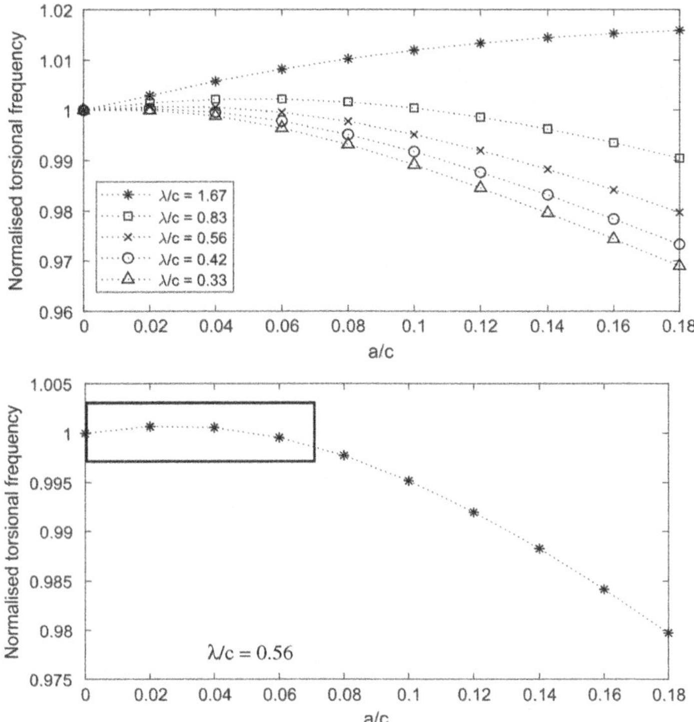

Fig. 11 Effect of increasing tubercle amplitude on first torsional frequency (zero phase lead)

3.2 Tubercles with π Phase Lead

The effects of varying tubercle amplitude and wavelength for tubercles of π phase lead are discussed in this section. Figure 14 shows the bending and torsional frequencies against amplitude of tubercles, respectively, for different tubercle wavelengths.

Regardless of the number of tubercles, both first bending and torsion exhibit a decrease in frequencies as tubercle amplitude is increased. The fall in frequency can be attributed to the trough close of the root that reduces chord length as tubercle amplitude is increased, thus reducing both bending and torsional frequencies. Similarly, the increased chord for the peak close to the tip leads to the effect of increased tip mass that lowers bending frequencies. This observation differs from the zero-phase tubercle arrangement in the previous section as the initial stiffening of the root with a peak is missing in the current π phase lead configuration. Another reason is due to the position of the C.G., which is moved towards the tip as tubercle amplitude is increased, as shown in Fig. 15. This increases flexibility and hence a decrease in the observed bending frequencies.

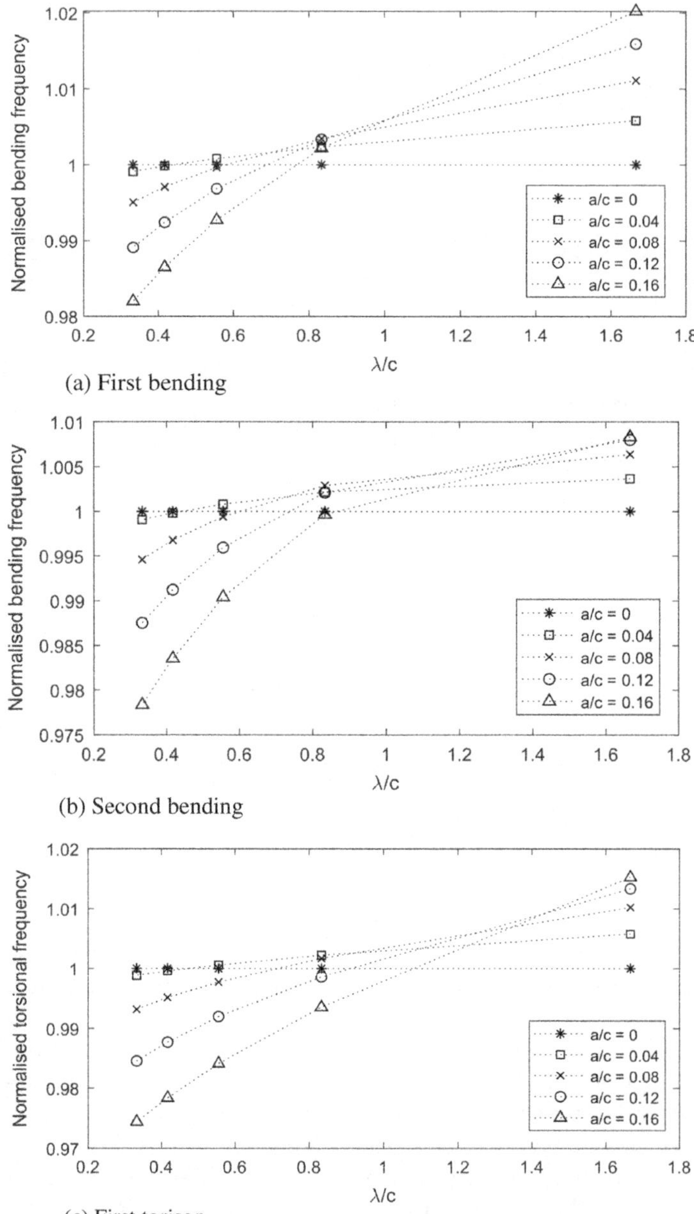

Fig. 12 Effect of increasing number of tubercles (reducing wavelength) on first, second bending and torsional frequencies (zero phase lead)

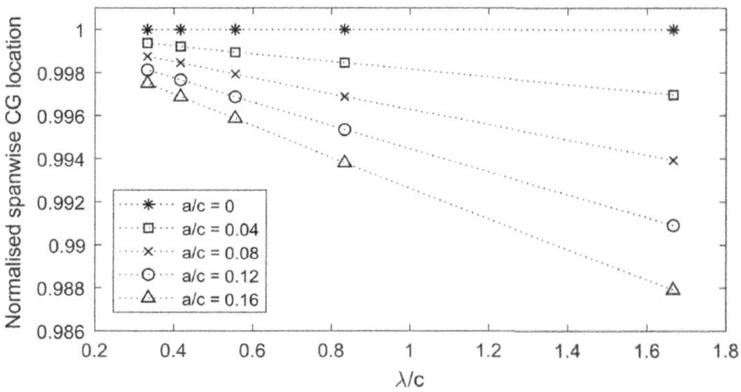

Fig. 13 Normalised spanwise C.G. location for increasing number of tubercles (reducing wavelength). Normalised with respect to mid-span (zero phase lead)

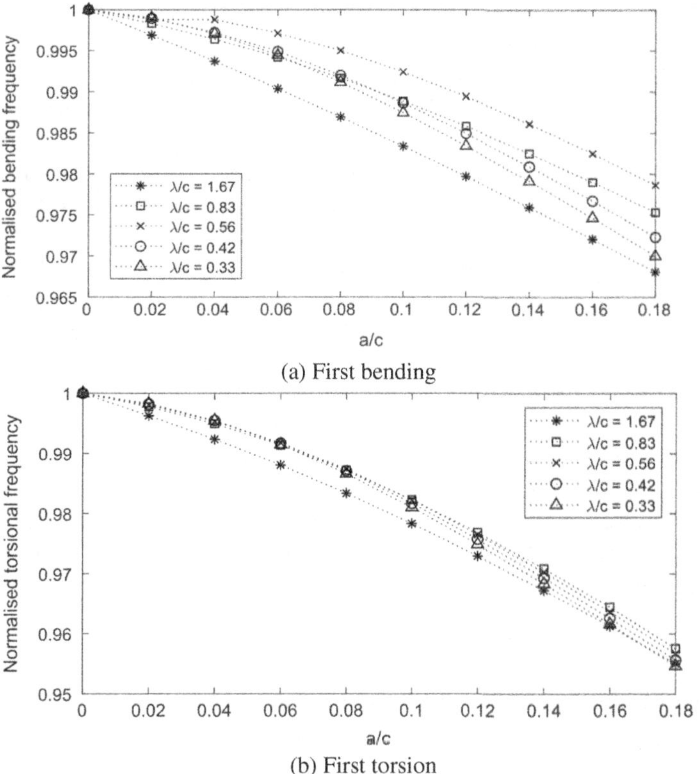

(a) First bending

(b) First torsion

Fig. 14 Effect of increasing tubercle amplitude on first bending frequency and first torsion frequency (π phase lead)

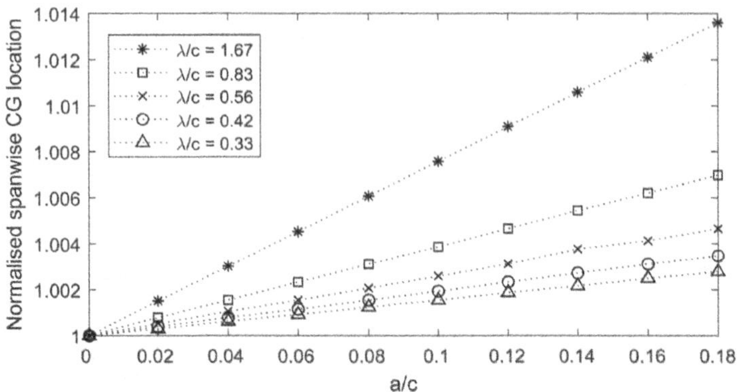

Fig. 15 Normalised spanwise C.G. location for increasing tubercle amplitude. Normalised with respect to mid-span (π phase lead)

As we increase the number of tubercles (wavelength reduced), the first bending frequency exhibits an initial increase up to when $\lambda/c = 0.6$ where it starts dropping after that. Initially, with two tubercles ($\lambda/c = 1.67$), larger tubercle amplitudes result in reduced chord close to the root in the π phase lead configuration. This leads to the initial lower bending frequencies for increased tubercle amplitude as observed in Fig. 16a. As tubercle wavelength is reduced to $\lambda/c = 0.6$ (increased to six tubercles), stability is improved as the spanwise C.G. location shifts from being closer to the tip towards the mid-chord in Fig. 17. However, as wavelength is reduced further (below $\lambda/c = 0.6$), this effect is offset as the peak and trough with increased and reduced chords are brought even closer to the tip and root, respectively, in the π phase lead configuration. This has the effect of increasing flexibility and hence the reduced frequencies observed.

The torsion frequency is little affected by changes in tubercle wavelength as shown in Fig. 16b. As tubercle amplitude is increased, due to the reduced chord at the trough regions, torsional stiffness is reduced. However, torsion is less affected by the number of tubercles as they are spread out evenly along the span. Additionally, the average chordwise C.G. position is unaffected by the number of tubercles for the same amplitude.

4 Effect of Tubercle Geometry on Aeroelastic Stability

4.1 Aeroelastic Formulation

Aeroelastic analysis requires a time-domain solution of a coupled system composed of a structural dynamics and an unsteady aerodynamics model. Here, the aeroelastic description is developed by Imperial College London to investigate the dynamics of

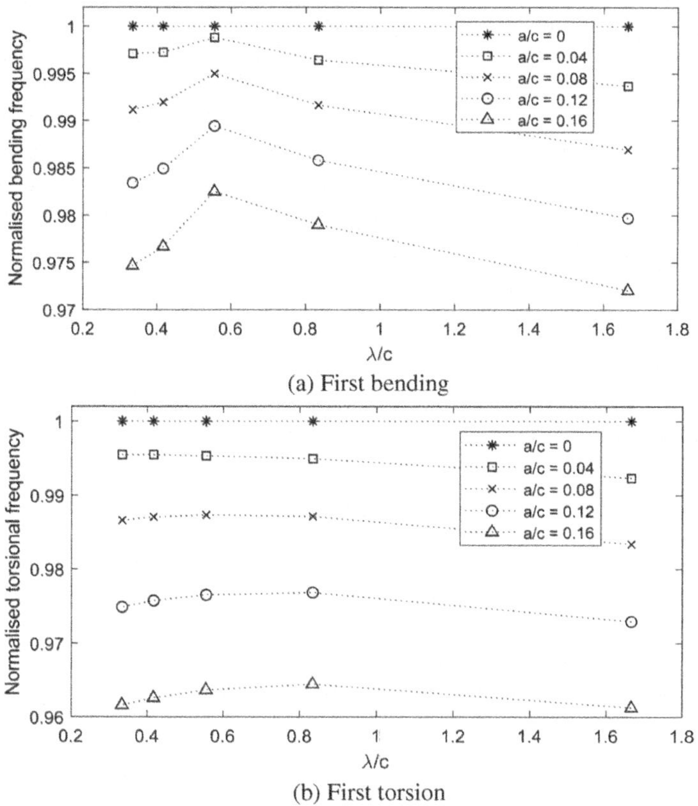

(a) First bending

(b) First torsion

Fig. 16 Effect of increasing number of tubercles (reducing wavelength) on first bending and torsion frequencies (π phase lead)

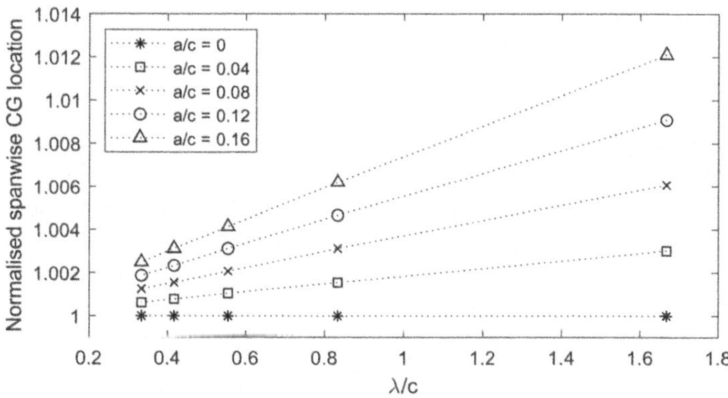

Fig. 17 Normalised spanwise C.G. location for increasing number of tubercles (reducing wavelength). Normalised with respect to mid-span (π phase lead)

flexible structures (Murua et al. 2012) for stability analysis, time-domain simulations and controls (Ng et al. 2014a, b; Ng et al. 2012; Van Parys et al. 2014).

The dynamics of the structure is described using composite beams, which is capable of accounting for large static and transient deformations. The nodal displacements and rotations are represented by the vector η from the finite element discretization. A cross-sectional homogenization approach is employed to represent 3-D structures as a 1-D beam description (Palacios and Cesnik 2005). For stability analysis, linearization is performed around an equilibrium η_0. The equations-of-motion (EoM) is expressed in incremental form as

$$M(\eta_0)\Delta\ddot{\eta} + K(\eta_0)\Delta\eta = \Delta Q_{ext}, \tag{1}$$

with M representing the mass matrix and K the stiffness matrix. Structural damping is neglected as it is small in comparison to aerodynamic damping. The vector Q_{ext} is used to introduce the aerodynamic forces.

The discrete-time unsteady vortex-lattice method (UVLM) is used to model the unsteady aerodynamics (Katz and Plotkin 2001) with a planar wake. The UVLM is valid for low-speed, attached and high Reynolds number flows. In the description of the UVLM, the fundamental solutions are represented using vortex rings that are arranged into lattices to represent the lifting surface. Neumann boundary conditions for zero normal flow velocity are subsequently imposed at each collocation point due to vortices as well as the dynamics of the lifting surface, represented by

$$A_{c,b}\Gamma_b^{n+1} + A_{c,w}\Gamma_w^{n+1} + w^{n+1} = 0, \tag{2}$$

where A_c is the aerodynamic influence coefficient resolved using the Biot-Savart law while Γ is the vortex circulations. The subscripts b and w represent the bound (lifting surface) and wake, respectively. The downwash w at collocation points is generated by the motion of the lifting surface and external disturbance. The wake vortex ring is shed from the TE and has a length of $U_\infty \times \Delta t$ where U_∞ is the freestream velocity and the time step $\Delta t = c/(U_\infty \times d)$ is determined by the UVLM discretization. c is the chord length while d is the number of bound chordwise panels (Ng et al. 2017b).

The fluid-structure coupling is illustrated as a block diagram in Fig. 18 with the flow of parameters. The structural degrees-of-freedom (dof) are mapped onto the blade lattice as downwash on the collocation point. On the other hand, aerodynamic forces are mapped to the structural nodes. Interested readers are referred to (Murua et al. (2012) and Ng et al. (2016b) for the detailed aeroelastic formulation.

To enable a state-space description of the unsteady aerodynamics in linear time-invariant form, a frozen wake in planar configuration is maintained (Murua et al. 2012). The continuous-time structural EoM are discretized using the Newmark-β method for subsequent coupling with the UVLM in discrete-time to arrive at a state-space description of the full aeroelastic system, represented by (Ng et al. 2017a)

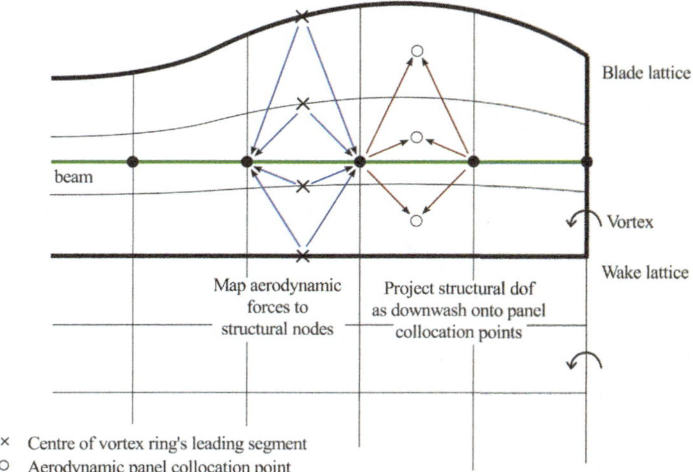

Fig. 18 Fluid-structure coupling

$$x^{n+1} = Ax^n + Bw$$
$$y^n = Cx^n + Dw,$$
(3)

where the state x contains aerodynamic circulations and structural dof. Stability of the system can be defined from the eigenvalues of matrix A.

4.2 Effect of LE Tubercles on the Aerodynamics in Pitch Motions

To appreciate the influence of LE tubercles on the non-stationary aerodynamics, the coefficient of lift without and with tubercles ($\lambda/c = 0.56$ and $a/c = 0.2$) undergoing harmonic motions of $\pm 10°$ angle of attack are plotted in Fig. 19 through the aerodynamics module in the abovementioned aeroelastic solver. Reduced frequencies of $k = 0.1$ and 0.4 are considered that correspond to unsteady flows that are mild and severe, respectively. $k = 0.4$ is in the proximity of the flutter reduced frequency of the Goland wing considered here. As shown, LE tubercles have relatively little effect on the unsteady aerodynamics pre-stall. On the contrary, the quasi-steady coefficient of lift is decreased by about 2% due to local spanwise velocity components generated by the wavy LE that decelerate flow to reduce LE suction, which improves stability margins (Bisplinghoff et al. 2013; Fung 2008; Theodorsen 1949; Theodorsen and Garrick 1940).

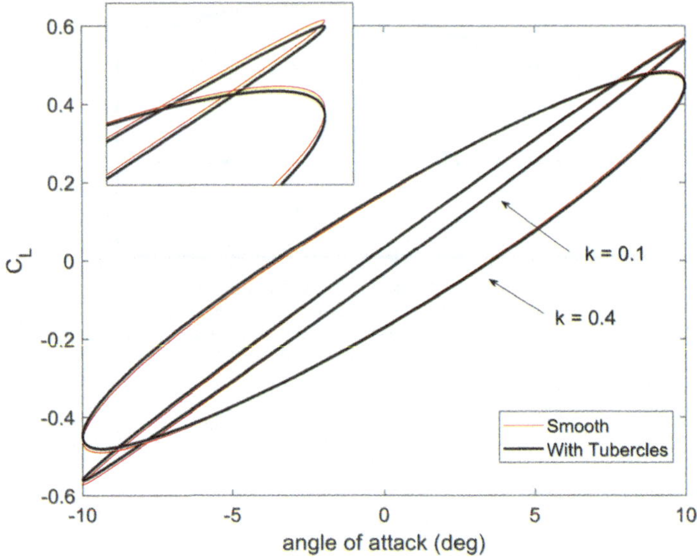

Fig. 19 Unsteady aerodynamics for lifting surface with and without tubercles ($\lambda/c = 0.56$ and a/c $= 0.2$). Lifting surface undergoes harmonic motion of k $= 0.1$ and 0.4

4.3 *Effect of Tubercles on Aeroelastic Stability*

As it has been observed, the unsteady aerodynamics with LE tubercles has the potential of increasing the stability margin due to reduced pre-stall lift (Sect. 4.2) while the structure changes may reduce the stability margin through increased flexibility (Sect. 3). Hence, it is critical to understand how these two mechanisms counteract to affect margins of stability when combined.

Two scenarios are considered—the effects of tubercles on aerodynamics only and on the coupled aeroelastic models. To account for tubercles in the structural module, the inertia and stiffness values in the composite beam are modified such that the natural frequencies of the lowest bending and torsional modes exhibit the same trends as that from Sect. 3. The variation of tubercle amplitudes on flutter speeds and frequencies are illustrated in Fig. 20. Here, the percentage variations with LE tubercles are computed against the baseline wing clean LE geometry.

Generally, with aerodynamics alone, up to 7% increase in flutter speeds can be observed as the amplitude of tubercle is increased to a/c $= 0.2$ with a uniform wavelength of $\lambda/c = 0.56$. This is the result of reduced pre-stall aerodynamic lift from the presence of LE tubercles. Additionally, there is a marginal frequency increase at flutter. When coupled in a full aeroelastic description, the reduction in natural frequencies lower the change in flutter speed to 5.5%, which is accompanied by a fall in the flutter frequency.

When the wavelength of tubercles is reduced from $\lambda/c = 0.83$ to $\lambda/c = 0.33$ (i.e. increasing number of tubercles from 4 to 10) with a/c $= 0.2$ in Fig. 21a, we observe

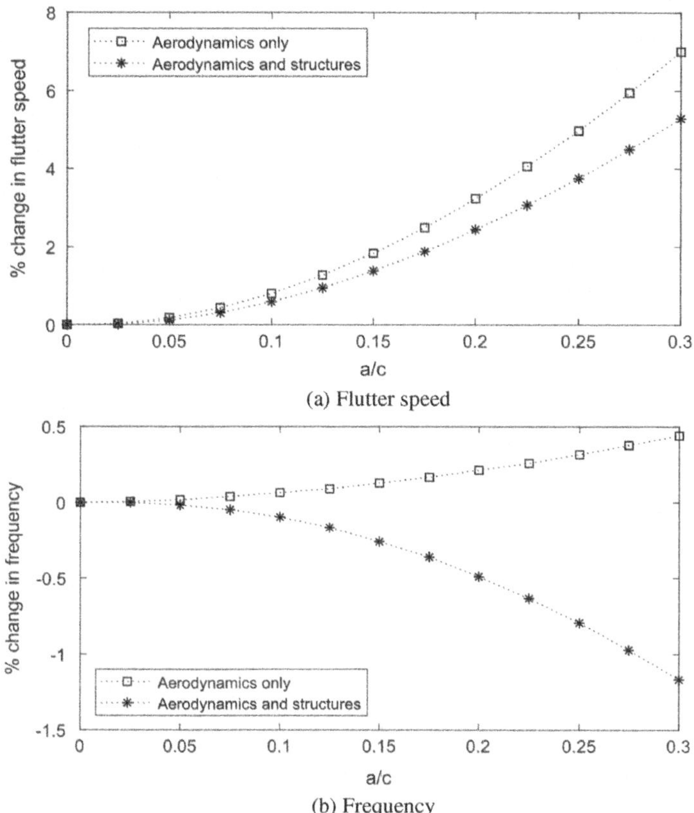

(a) Flutter speed

(b) Frequency

Fig. 20 Increasing tubercle amplitude on flutter speed and frequencies ($\lambda/c = 0.56$)

an increase in flutter speeds of 5.5% with aerodynamics alone and up to 5% for full aeroelastic description. The frequency at flutter is higher in the former and lower for the latter in Fig. 21b.

5 Spanwise Variation of Tubercle Geometry

Localised effects can be observed when tubercle wavelengths and amplitudes are varied spanwise. Table 2 illustrates the change in flutter speeds for spanwise variation of tubercle wavelength and amplitude on the Goland wing.

Configuration A1 is without LE tubercles and serve as the baseline for comparisons. Configurations A2 and A3 are equipped with LE tubercle where the amplitudes and wavelength are uniformly distributed spanwise. In configuration B1, the tubercle wavelength is modified spanwise from $\lambda_r/c = 0.70$ at the root to $\lambda_t/c = 0.33$ at the tip (i.e. wing tip has more tubercles) while keeping an amplitude of $a/c = 0.2$ uniformly

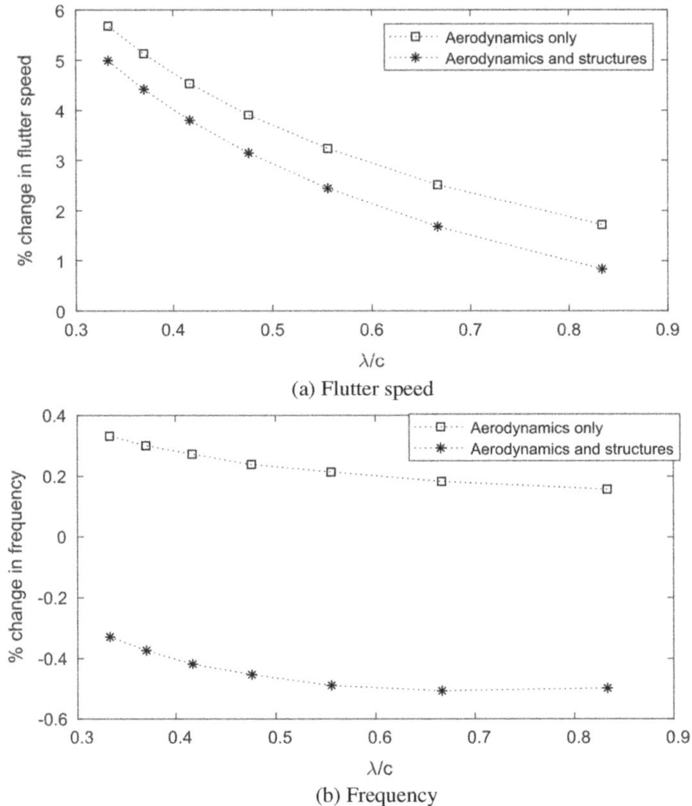

(a) Flutter speed

(b) Frequency

Fig. 21 Increasing tubercle wavelength on flutter speed and frequency (a/c = 0.2)

across the span. This causes the flutter speed to increase by 2.9%. Conversely, when more tubercles are positioned at the root (configuration B2), the increase in flutter speed is less significant at 1.7%. In configurations C1 and C2 where tubercle amplitudes are varied spanwise while maintaining a wavelength that is uniform, flutter speed are higher when the amplitude of tubercles are larger towards the tip. From the aeroelastic standpoint, smaller displacements/rotations are experienced at the wing root as compared to the wing tip. Consequently, the effects on aerodynamics and stability are less pronounced when there is a larger concentration of tubercles or larger tubercle amplitudes towards the wing root.

Table 2 Percentage variation in flutter speeds for different LE tubercle amplitudes and wavelengths

Configuration		a_r/c	a_t/c	λ_r/c	λ_t/c	% Δ flutter speed
A1 (baseline)		0.0	0.0	0.0	0.0	–
A2		0.1	0.1	0.56	0.56	0.59
A3		0.2	0.2	0.56	0.56	2.44
B1		0.2	0.2	0.70	0.33	2.94
B2		0.2	0.2	0.33	0.70	1.65
C1		0.0	0.3	0.56	0.56	2.31
C2		0.3	0.0	0.56	0.56	0.99

All configurations are designed with six tubercles with the same planform area
Subscripts r and t are for root and tip, respectively

6 Sweptback Configurations

Common lifting surfaces possess a certain degree of sweep and here, we discuss using swept-back configurations of the Goland wing (Fig. 22) to illustrate how LE tubercles affect aeroelastic stability under conditions with spanwise flow.

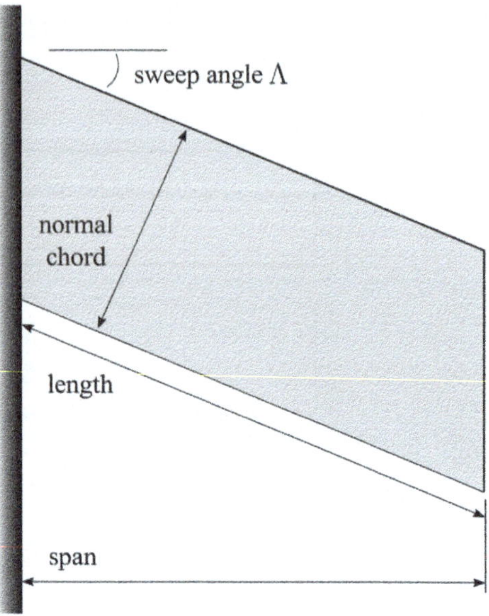

Fig. 22 Definition of a swept wing configuration

There are two ways for sweep—shear-swept (SS) and constant length-to-chord ratio (CLC) (Babister 1950; Barmby et al. 1951; Molyneux 1950) as shown in Fig. 23. In the SS model, cross-sections remain parallel to freestream where the normal chord is reduced while keeping the wing span and area unchanged. As such, the aspect ratio is maintained. In the CLC model, the wing is revolved while both ends are adjusted to be in line with the freestream. Other parameters including chord, length of wing and area remain unaffected.

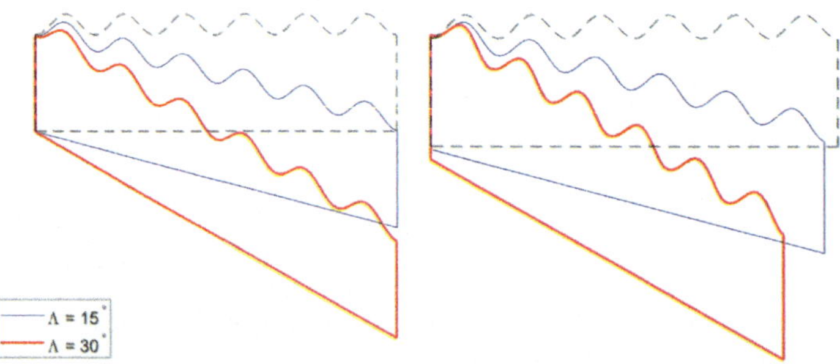

Fig. 23 Swept wings with of SS (left) and CLC (right) swept model

6.1 Effect of Tubercle Amplitude

In accounting for LE tubercles on swept wings, the amplitude and wavelength are parameterized with respect to the chord perpendicular to the LE. As we increase the sweep angle in both the SS and CLC models, tubercles of larger amplitudes have less effect on flutter speeds (Fig. 24). At the wing sweep of 15°, flutter speed is increased by a maximum of 5% in contrast to 3.5% for wing sweep of 45° in both the SS and CLC configurations.

Increasing spanwise effects resulting from sweep is the main contribution to the reduced effect of LE tubercles on flutter speed. In the SS wing, tubercle wavelength is increased by the longer wing span for same number of tubercles, thereby weakening the efficacy of the tubercles. In the CLC wing, the effective tubercle amplitude is

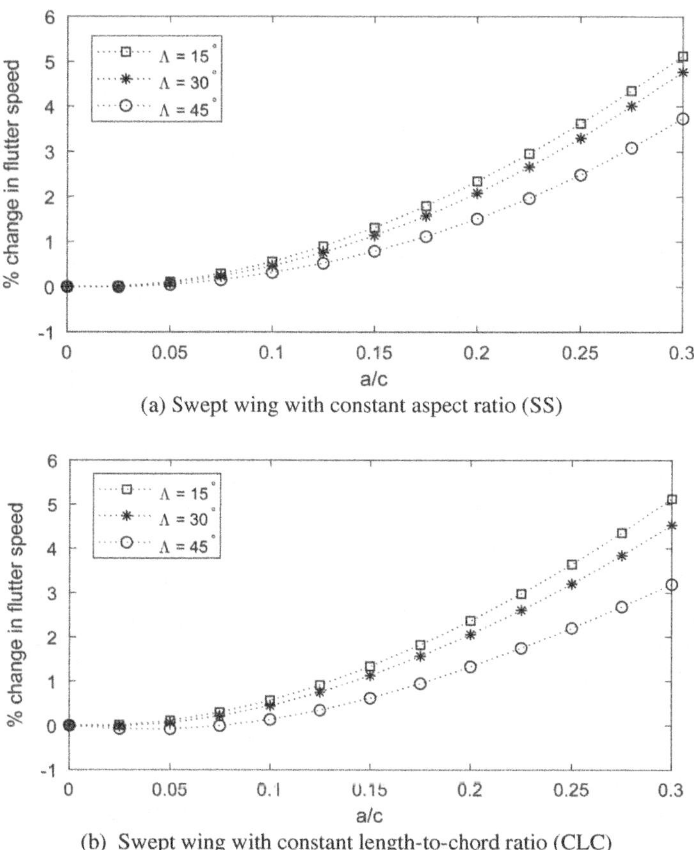

Fig. 24 Effects of increasing tubercle amplitude on wing flutter speeds for different swept angles. SS and CLC models with six LE tubercles

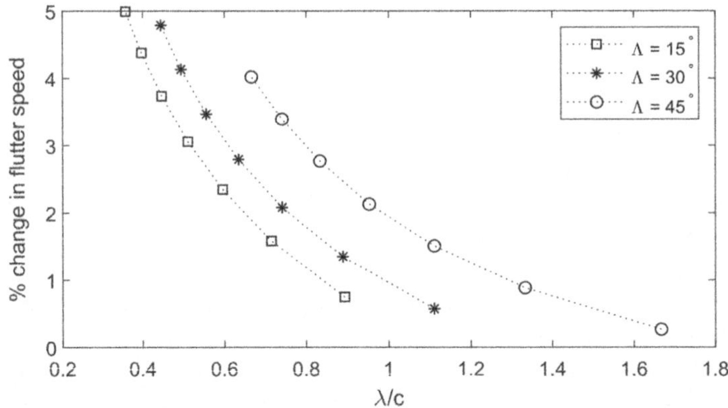

Fig. 25 Effect of increasing number of tubercles (reducing wavelength) on flutter speeds for wings with different angles of sweep. SS model and a/c = 0.2

reduced in relation to the swept chord, which is parallel to freestream. This has the effect of weakening LE tubercles on flutter speeds.

6.2 *Effect of Tubercle Wavelength*

When LE tubercles are increased from four to ten (while keeping amplitude constant at a/c = 0.2), both swept wing configurations observe increase in flutter speeds. In the SS wing in Fig. 25, as swept angle is increased, having more LE tubercles have lesser effect on flutter speeds. As observed, the flutter speed is increased by a maximum of 5% for the wing that is swept at 15°, in contrast to 4% for the wing that is swept at 45°. This can be due to the longer wing length that results in larger tubercle wavelength.

In the CLC wing in Fig. 26, the effect of sweep on flutter speeds become increasingly similar for larger number of LE tubercles. With four LE tubercles ($\lambda/c = 0.83$), wing sweep of 15° has larger effect on flutter speeds than the wing with 45° sweep. However, differences between the swept wings are lessened with increased number of tubercles.

7 Conclusion

Bio-inspired engineering opens opportunities to improving aerodynamics but comprehensive analysis is necessary to investigate structural responses and stability margins for a given amount of lift. While the discussions in this chapter are restricted to flat plate configurations, it provides insights to the potential impact on structural

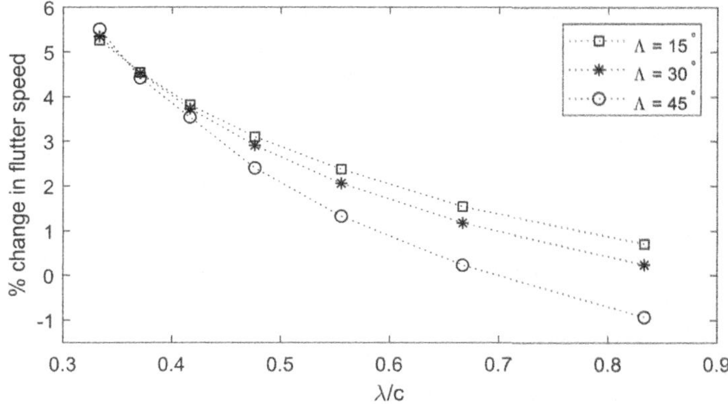

Fig. 26 Effect of increasing number of tubercles (reducing wavelength) on flutter speeds for wings with different angles of sweep. CLC model and a/c = 0.2

responses of lifting surfaces modified with LE tubercles that warrants a comprehensive analysis on top of pure aerodynamic design. On actual wings or blades with airfoil cross-sections, modifications to internal layout and topological optimization is also required. Finite element software should be used for detailed static and dynamic structural analysis while efficient aeroelastic tools serve as efficient means for stability analysis.

With LE tubercles, bending and torsional frequencies are slightly lower due to the combined effects of spanwise stiffness and inertia variations from mass redistribution and spanwise chord length variations. Aerodynamics and structures resulting from LE tubercles have counteracting effects on the margins of stability. The former is responsible for increasing stability margins from lower lift and the latter decreases stability margins as a result of lower natural frequencies. Combined in an aeroelastic formulation, aerodynamics dominates and flutter speeds are increased. Additionally, having more LE tubercles towards the root of the wing have lower influence on stability margins. On sweptback wings, stability margins are less affected by LE tubercles as the angle of sweep is increased.

References

ANSYS® Academic Research Mechanical, Release 18.1. (n.d.)

Babister AW (1950) Flutter and divergence of sweptback and sweptforward wings. College of Aeronautics Cranfield

Barmby JG, Cunningham HJ, Garrick IE (1951) Study of effects of sweep on the flutter of cantilever wings

Bisplinghoff RL, Ashley H, Halfman RL (2013) Aeroelasticity. Courier Corporation

Bolzon MD, Kelso RM, Arjomandi M (2016) Parametric study of the effects of a tubercle's geometry on wing performance through the use of the lifting-line theory. In: 54th AIAA Aerospace Sciences Meeting. American Institute of Aeronautics and Astronautics, Reston, Virginia

Corsini A, Delibra G, Sheard AG (2013) On the role of leading-edge bumps in the control of stall onset in axial fan blades. J Fluids Eng 135(8):81104

Fish FE (2009) Biomimetics: determining engineering opportunities from nature. In: Proceedings of SPIE Conference, SPIE, vol 7401, p 740109

Fung YC (2008) An introduction to the theory of aeroelasticity. Courier Dover Publications

Goland M (1945) The flutter of a uniform cantilever wing. J Appl Mech Trans ASME 12(4):A197–A208

Goruney T, Rockwell D (2009) Flow past a delta wing with a sinusoidal leading edge: near-surface topology and flow structure. Exp Fluids 47(2):321–331

Hansen KL (2012) Effect of leading edge tubercles on airfoil performance. Ph.D. Thesis, University of Adelaide

Hansen KL, Kelso RM, Dally BB (2011) Performance variations of leading-edge tubercles for distinct airfoil profiles. AIAA J 49(1):185–194

Hansen K, Kelso R, Doolan C (2012) Reduction of flow induced airfoil tonal noise using leading edge sinusoidal modifications. Acoust Australia 40(3):172–177

Howle LE (2009) Whalepower Wenvor blade: a report on the efficiency of a whalepower corp. 5 meter prototype wind turbine blade. Technical report, Bellequant Engineering, PLLC

Katz J, Plotkin A (2001) Low speed aerodynamics, 2nd edn. Cambridge University Press, Cambridge

Malipeddi AK, Mahmoudnejad N, Hoffmann KA (2012) Numerical analysis of effects of leading-edge protuberances on aircraft wing performance. J Aircraft 49(5):1336–1344

Miklosovic DS, Murray MM, Howle LE, Fish FE (2004) Leading-edge tubercles delay stall on humpback whale (*Megaptera novaeangliae*) flippers. Phys Fluids 16(5):L39–L42

Molyneux WG (1950) The flutter of swept and unswept wings with fixed-root conditions. Citeseer

Murua J, Palacios R, Graham JMR (2012) Applications of the unsteady vortex-lattice method in aircraft aeroelasticity and flight dynamics. Prog Aerosp Sci

Ng BF, Palacios R, Graham JMR, Kerrigan EC (2012) Robust control synthesis for gust load alleviation from large aeroelastic models with relaxation of spatial discretisation. In: Proceedings of European wind energy conference and exhibition. Copenhagen, Denmark

Ng BF, Hesse H, Palacios R, Graham JMR, Kerrigan EC (2014a) Efficient aeroservoelastic modeling and control using trailing-edge flaps of wind turbines. In: Proceedings of UKACC International Conference on Control, Loughborough, United Kingdom

Ng BF, Hesse H, Palacios R, Graham JMR, Kerrigan EC (2014b) Model-based aeroservoelastic design and load alleviation of large wind turbines. In: Proceedings of 32nd ASME wind energy symposium, AIAA SciTech. National Harbor, Maryland, USA

Ng BF, Hesse H, Palacios R, Graham JMR, Kerrigan EC (2015) Aeroservoelastic state-space vortex lattice modeling and load alleviation of wind turbine blades. Wind Energy 18(7):1317–1331

Ng BF, New TH, Palacios R (2016a) Effects of leading-edge tubercles on wing flutter speeds. Bioinspir Biomim 11:3

Ng BF, Palacios R, Graham JMR, Kerrigan EC, Hesse H (2016b) Aerodynamic load control in horizontal axis wind turbines with combined aeroelastic tailoring and trailing-edge flaps. Wind Energy 19(2):243–263

Ng BF, New TH, Palacios R (2017a) Bio-inspired leading-edge tubercles to improve fatigue life in horizontal axis wind turbine blades. In: 35th wind energy symposium, p 1381

Ng BF, Palacios R, Graham JMR (2017b) Model-based aeroelastic analysis and blade load alleviation of offshore wind turbines. Int J Control 90(1):15–36

Ozen CA, Rockwell D (2010) Control of vortical structures on a flapping wing via a sinusoidal leading-edge. Phys Fluids 22(2):21701

Palacios R, Cesnik C (2005) Cross-sectional analysis of non-homogeneous anisotropic active slender structures. AIAA J 43(12)

Theodorsen T (1949) General theory of aerodynamic instability and the mechanism of flutter. National Advisory Committee for Aeronautics Report No. 496

Theodorsen T, Garrick IE (1940) Mechanism of flutter a theoretical and experimental investigation of the flutter problem. National Advisory Committee for Aeronautics Report No. 685

Van Nierop EA, Alben S, Brenner MP (2008) How bumps on whale flippers delay stall: an aerodynamic model. Phys Rev Lett 100(5):54502

Van Parys BPG, Ng BF, Goulart PJ, Palacios R (2014) Optimal control for load alleviation in wind turbines. In: Proceedings of 32nd ASME wind energy symposium, AIAA SciTech. National Harbor, Maryland, USA

Wang Z, Chen PC, Liu DD, Mook DT, Patil MJ (2006) Time domain nonlinear aeroelastic analysis for hale wings. In: 47th AIAA/ASME/ASCE/AHS/ASC structures, structural dynamics, and materials conference 14th AIAA/ASME/AHS adaptive structures conference 7th, p 1640

Watts P, Fish FE (2001) The influence of passive, leading edge tubercles on wing performance. In: Proceedings of twelfth international symposium unmanned untethered submersible Technology, Autonomous. Undersea Systems Institute, Durham New Hampshire

Wei Z, New TH, Cui YD (2015) An experimental study on flow separation control of hydrofoils with leading-edge tubercles at low Reynolds number. Ocean Eng 108:336–349

Zhang M, Frendi A (2016) Effect of airfoil leading edge waviness on flow structures and noise. Int J Numer Meth Heat Fluid Flow 26(6):1821–1842

Index

© Springer Nature Switzerland AG 2020

D. T. H. New and B. F. Ng (eds.), *Flow Control Through Bio-inspired Leading-Edge Tubercles*, https://doi.org/10.1007/978-3-030-23792-9

The manufacturer's authorised representative in the EU is Springer
Nature Customer Service Centre GmbH, Europaplatz 3, 69115 Heidelberg,
Germany. If you have any concerns regarding our products, please
contact ProductSafety@springernature.com

Printed and bound by CPI Group (UK) Ltd, Croydon, CR0 4YY
28/11/2025
02007692-0001